Tended Stones

Cote D'Ivore Blueprints

Benjamin (Allan)

Schmidt

---Page 1--------------

1

. In2005,A scientist,MarySchweitzer`steam,publishedinthejournal Science newsthat

rai s edcommoti ni sci e nti f i c

ci r cl e s.Herteam,afterdi sol v i n ganewl y foundTyrannosaurusRex`sbonei asol u ti o n thatcompl e tel y di s i n tegratedthe

fossi l matter,di s coveredactualbl odvessel ,bonematri x andosteocytes 3 . 530. StephanCulbrethlivesinElGranada,California,andformorethanadecade

studi e sfi n dsthatretai n tracesof

di n osaurhabi at;stoneswhi c hhehi ms el f begantocal "MysteryStones",andIgave

them thename Brass Stones A , and Tended stones. B

531. The rocks carryresemblancetofossilorgans.Nomatterhow thismaysoundif

comparedtotheMarySchwei zer`s di s covery,

There are ways and possibilities how to approach drawing and in the following text, I will be concerned about the stilistic manner for the drawing of Palaeontological finds. There is a wide history of drawing and engraving Palaeontology, and the earliest known traces of this craft reach to the end of the 19th century, which comprised great personages for the upheaval of the classical

palaeontological trends and novelties in education. The first known forms of drawing of Palaeontological

finds comprise etchings (mostly Aquatint), copper plate and wood cut.

Aquatint is a teqnique mostly performed on a copper plate or a brass plate with the help of a tar polish and a needle for necessary engraving. By submerging the done plate into a mild acid, the 4

necessary grooves hollow out into the plate into which a print color is then

implemented. Copper plate is performed only with the use of an engarving needle and subsequent use of print colour, while wood-cut is usualy cut into mild and soft wood with the help of small wood chistles. In this book, I will be more interested in drawing and possible prognostics of modern use of digital drawing for etimological and taxonomical purposes.

Drawing is a wonderful pass time for leasurly days and can become your good friend when practiced with easiness and

diligence. There is a big difference between drawing and painting.

Drawing is usually performed with a pencle, occassionaly even coloured pencils (for the drawing of fish on research boats), charcoal or ink. There is a good way how to study shape and structure by drawing 5 still-lifes, often under the supervision of a lectore. The possibilities of studying drawing for the sake of detailed drawing of palaeontological finds can result in Museum career, or at least obtaining a diploma for bettering your further career plans. There is a wide probability that if you reach for a stigmatic life on research programmes, you may find your life inspiering enough to write a book, or endulge in a good name. There are differen ways how to taxonomicaly measure a find and here I'm more

concerned about the way digital pen looks and its prognostics for a daily use in drawing Palaeontology. Beetles in palaeontology can be drawn in

easified half-sections or whole containing colours. There are different ways how to approach 6

shading and its possible results should be workd with only in a small portion. John Demoor's stones on the beaches – the possibility of a fossilization of whale is an interesting way how to resolve the

mystery of Tended stones. Tended stone is a stone which underwentoutside cause for its origination. The possibility of falling pray is not excluded and the subsequent fossilization or calcification in Deep Freez theory took place in close beach environment. The possibility of harvesting such stones out of querries in not excluded.

Niamh Noonan, county Kerry – Architectonical blueprint Katie Burgess and her daughter Scorcha are very nice friends.

Excavations 1. of paleolith in Prague subarb, Excavations in France of Neolith, and paleolith,

Excavations in Kerry and Italian Alps. Geological anomalies.

---Page 2--------------

Solnhofen lithographic stone for making prints, just as for example Alfons Mucha. Print was done on an old printing machine.

Small crocodille wooden toys, like the name of a known digeridoo band (for charity shop). Benjamin Schmidt, left above. Meditating people in the street right above.

Plan for Raiderpublishing in Europe after Europian Identity (Twopeople)

Bar J.J. Murphy is open daily at 7 o ' clock in the morning to get breakfast, lunch o ran evening fun for students or holiday makers, erasms, or interesteers in deepening and broadening your knowledge on an international bettering of voccation knowledge. Prague Centre.

Visit galleries, meetings and bars in your free time for it coresponds with cultural knowledge where you can rest and asses your further bettering voccation plans.

Cultural Identity was published both in Europe and Chicago (America) for your knowledge of Europian Union and Shengen (Your opportunities there, career and bettering).

---Page 4--------------

Plan for Raiderpublishing in Europe after Europian Identity (Twopeople)

Study the making of furniture (Rocco furniture, Ludvig 14., etc. Polishing methods, kunshistoric approbation etc.).

--Page 5--------------

Plan for Raiderpublishing in Europe after Europian Identity (Twopeople)

Turn a carver or a master of craft. Better your knowledge on international studies or vacation.

Create accounts for accounting your paraphernelia on your way back, or stey to transfer your work abroad in a legal way.

---Page 7--------------

Plan for Raiderpublishing in Europe after Europian Identity (Twopeople)

Lithographic printing teqniques are nowadays done and performed on high quality large machines which work without flaw. Better your knowlede of offset print and old teqniques in an arrey of printig studios all over Europe where people nowadays can communicate in English (Europian English Standard)

Plan for Raiderpublishing in Europe after Europian Identity (Twopeople)

TWOPEOPL
 E

TWOPEOPLE

Benjamin Schmidt

Plan for Raiderpublishing in Europe after Europian Identity (Twopeople)

Cal c i f i c ati o n is the process in which calcium salts build up in soft tissue, causing it to harden. Calcifications may be classified on whether there is mineral balance or not, and the location of the calcification.

Calcification Table:

- ironisation/ opalisation mostly veins

Soft Tissue - Calcium (calcification) - Quarts - Talc fat - Feldspar muscles - Schist (silificated)

We are trying to expound on the fact that not only it is possible for remnants of

soft-tissue to get preserved on fossil-bones 1 , but we would like to show this phenomenon in a completely new light; bringing down our evidence based on collective observation, 2 and field research, that whole organs of prehistoric life have been found preserved up to the present, and hereby illuminate a scenery that I began to call a 'Brass-

These attributes (as I call such palaeontological phenomena collectively) are usually, if not always, found in groups. They seem to present a nest, a probable migratory roosting place, and a stereotypical ritualistic behaviour of Miocene birds.

I claim the possibility of an unmarked heavy glaciation (Ice age) during the Mi o ceneEpoch [Terti a ry boundary]. In other words, the possibility of an unmarked rapid temperature fall during the early Miocene and seasonal arising of secluded environments that respectfully came into existence in these extreme weather conditions.

I claim a trait of a ritualistic behaviour in this type of bird: building of nests, calcificated organs from marine [mammalian] creatures. A possible cuckoo behaviour. [Raising dinosaur eggs for the purpose of storing food is not excluded]; presupposing remaining dinosaur species of small stature.

---Page 10-------------

11

Resuscitation and drawing of a Celtic Clasp from around the 1st century B.C. Europian Celtic Heretage. Drawing by Bejamin Schmidt (replica).

Excavations of paleolith in Prague subarb, Excavations in France of Neolith, and paleolith, Excavations in Kerry and Italian Alps. Geological anomalies.

--Page 11-------------

Ecological and economical estates in modern neo-Victorian styles. Architecture (proksprikril).

I correct Sorcha - name. Bhante Y. Wimala, Ashin Ottama.

Rest of the index

Bhante Y. Wimala(2001) Lessons of the Lotus - Sri lanka print, 25p.

Bonnie Greenwell(1995) Energies of Tranformation, A guide to the Kundalini process , Delhi, 105p.

Alan Charig(1992) A new look at Dinosaurs , Great Britain press, 125p. Vojtech Novotny (2005) Notebooks from New Guinea , BBC,153p.

Owen R. (1855), Lectures on the Comparative anatomy and Physiology of Ivertebrate animals, Delivered at the Royal collage of Surgeons . Longman, London, 689pp.

Dawson, J.W. (1888) The Geological History of plants . Kegan Paul, Trech & Co., London -and Callow R.H.T. (2007). Changes in the patterns of Phosphatic preservation across the Protozoic-

Cambrian transition Memoirs of the Asociation of the Australasian Palaeontologists .

--Page 13-------------

Remaining sentences

We visit the Zoo in midland India and meet a Gorilla with a funny name Bopa.

There is not much to do except to drink water in the hot weather.

Italian Rome is a vast city full of beautiful architecture and pleasant people who

always try to help you with a smettering of English.

They show you directions and laugh.

Pompeys belong to one of the most important National Heretages in Italy and can

represent future beautiful memories.

If you decide for a break in Varanasi ghats to visit a stoopa, you may find interesting

that Jalahar word for a doctor belongs still almost to the mundane.

The Trans-Oriental Express is still a certain bet to travel to the East from Europe in

about fourteen days.

The Zakarpat Express is also still functionate in Alps and the slow-down into the

steep mountains belong to one of the most incredible adventures in life.

The Buddhist monastery Santacittarama in Italy houses several monks with a cook. The food is prepared daily at 11:00 o'clock and represents the only dish of the day. Buddhist monasteries in Europe are open to newcomers and participants who freely

fluctuate through them just for a couple of days.

If you wish to visit Denmark and do not feel like flying, you may also sail from

Germany.

French people like food, and food has always been a part of their culture. Larochell belongs to one of the most beautiful summer resorts in Europe, just as Niess, etc.. If you wish to go north of India, you may as welll visit Dalailama in Dharamsala. Sauriel park at Kleinvelka is a vast complex of a park imbeded in a forest and is

worth visiting.

There are dinosaur museums in France of a relative largness and far succeed the rest

of Europe.

The French person Charles Lemark was the first person who coined the term Fossil. You may watch Soup opera in Every Europian country, or study on a mutual agreement.

1.

Remaining sentences from Africa

for browsing

We cook almost without the help of cooking gear. The weather is occassionally very

strong and we are trying to seclude ourselves into a shelter.

Logical stipuly according which we can judge the age of a layer with basical tools can lead us into understanding of fossilization.

We were slightly scared if we could stay in the monastery Santacittarama, and we did not have problems. The welcome was incredible.

Wheather the finds are found in Ethiopia, or secluded place for possible finds in Europe is an interesting prodilema for interest. Lay practitioner reads a book during a lunchbreak.

People can distinguish between Pali read text and Hindi speech in India - Jalahar. Mantras are usually recited in Pali.

I break a bar of chocolate and wonder that it does not melt. We visit Africa only for several days.

Diying - washing clothes lying - on bed with newspapers should be properly

corrected. Possibly some mistakes.

With deep gartitude I thank you for helping me finish up a very good journal. Everything was made safe and my professors helped me with so much. I conclude that we wish just harmony and are sending the best we currently have. Its possible that we might one day surprise yet at least with an article.

Fictitious stories Julian. I correct Jon Cohen Mediocre science. Benjamin Schmidt

1.

We are looking for people to take part in a palaeontological research.

Co. Clare, close to Tralee Spring 2015 (possibility of payment).

All information and a map will be on our website

Please send your (CV) and a motivation letter at woodenclaw@email.cz

+420/608268129 a list of chosen people will be informed in time

We are also looking for a driver.

Nature must be kept and preserved. We strive to show people how to

understand natural phenomena such as they were, their affinity to life, and the

life they pertained. Volunteers welcomed for

Introduction to: palaeontological research autumn preparations

- Jurassic /Miocene - taxonomy - geology

- botanology

Under the auspices of ....

school visits welcome

Allen Petterson & Co.

Http://gallerymarvels.com ... all information,.. from autumn 2014

we lecture on: tools of art dinosaur park at Kleinvelka in the track of John Walcott Vaccination

(we strongly reccommend a good mood and a will for supperstition)

Mendatory vaccination for hapattites A.

Portrait - Africa- Benjamin Schmidt - Interesting visit in Santacittarama in Italy. Pierre stays in the hotel to work out his education. Me and Pat are painting three kuties (meditational cubicles). The winter is about 15-20 degrees. Ajhan Visuparo is going to be closed in for a year and several days.

I can be anytime taken downtown in a monstrous jeep driven by a 22 year Robie (from Rome). We are building in a forest a huge meditational structure. Ye Hindi he? Nadi kaha Pare? You have to be careful for glass-foam plates and Monks with versatil moods.

---Page 17-------------

Pencil case for drawing Archaeology and palaeontology should contain
pencils (colored),rubber, glove, sharpener, ink, and a rag. A brush and
pad of papers.

---Page 18-------------

Tended stones - Blueprints from Papua New Guinea

Palaeontological Centre and Shimpanzee reserve blueprint. Stairs can leed to a round window classical in 1910 Papua New Guinea structures. Shop, or pottery shop.

Betterments with statues (Colonial style, Far East) Large water gurgling statues. Palaeontological centre.

--Page 20-------------

I

--Page 21-------------

It is said, that when you submerge pearls into a glass of water, only the true ones will remain dry. With rubies, it's the same. I'm Julian, and this is my daughter Cecil! I left my second wife in Calcutta and am now traveling north of India, in search of a jewel which the local inhabitants call The Miracle of the East. Hold me thumbs, this is going to be a ride!

Me and Cecil were waiting at the platform 5. The train that was supposed to lead us to Varanasi was already in. The Train Station was full of people. Literally crowded with tourists and small tradesmen, waiting for their allotted departure time.

..

II

---Page 22-------------

III

--Page 23-------------

Allen Petterson (Benjamin Schmidt)

IV

--Page 24-------------

V
---Page 25--------------

What are we waiting for, Daddy? Asked Cecil. I was supposed to meet doctor Forlore here, my darling! I said. Then he appeared. The steam of the train if front of us for a while sheltered our view. Here, Julian! Said doctor Forlore, in an unusually good humour. I packed you some good things. Don't forget to eat the sandwiches! They are on top. You are going to be in the business class. Don't stray on the train, it may be dangerous up North.

Doctor Forlore was giving us instructions, when I noticed an exceptionally beautiful woman, probably from Dargeeling, having a heated dispute with an older man in an expensive suit. He was holding his breast watch and seemed little inclined to listen to her reprimands.

Come! The train is leaving! I shouted at my straying daughter. And we were obliged to enter the train already on the move. All to the reprisals of the local guards, who were impatiently waving at people to orderly board the steaming monstrum.

--

And this is your coupe, Sir, Lady! The dinner's at seven, in the wagon 3. Thank you, I said to the youngish man, possibly in his twenties, and gave

VI

--Page 26-------------
----

him a two dollar tip. He seemed happy enough with it as to offer me extra sheets and a heat up mattress for my daughter, which he was supposed to bring, later on.

Me and my daughter were playing a board game we found in a cupboard, next to where our luggage was. It carried a strange inscription in Hindi, which meant something like: There you go again! We set up the figures according to a game we knew from home, and each time, when one of us thought he was winning, one would say: There you go, again!

Suddenly, we heard a noise from the corridor. Someone was talking to the train guard. Who do you think is going to win the next Birmingham final? He was apparently English. I couldn't guess, whether he was tall, fat, or slender. I think the Partisons might, answered the train guard.

I was surprised to hear such good English out of train guards, though this time, it was a different man than the one who was supposed to bring us the extras. Nice weather up here! Carried on the Englishman. Indeed, sir! We are at the end of spring. It will be only better and better.

Better you play! Said Cecil.

I put my figure instead of one of her's, and said:

Here you go, again!

VII

--Page 27------------

--

The dishes were served on a small trolley, by a boy possibly of fifteen. Cecil ate her portion of mashed green peas with interest. Next to us, across the aisle, were a lady of around sixty with her husband Charles. As I gleaned from the conversation, they were old English ex-patriots with firm roots in Bombay. She was called Miriam. An unusual name for a woman of her age, more usual for the Colonial India. They traveled North for business. For what kind of business, I couldn't guess. When I looked Cecil over the shoulder, there was a family of three, with a son David. They looked western, but their English was broken. After them, there was a small, fat man with a pince-nez and scrupulous manners, indulged in a conversation with a woman of around forty. Plainly dressed, with a pearl necklace, she looked like the daughter of a rich ambassador. On the left, across the aisle off these two, which I later found to be called Pierre and Charlotte, was sitting the enigmatic beauty from the East. She was so beautiful I could barely rest my gaze at her. I thought of my wife and averted my look to Cecil. Do you like the mashed peas? I asked. Yes, Daddy, they are delicious!. I hardly ate.

VIII

I rose from our table. David from the next table came to meet me. My father says, he would like to introduce himself to you. Certainly! I said

I'm Peter, meet my wife Juliet! Said the man. Are you French? I asked. My wife, yes! I was raised in Brussels. Did, you like the food? I asked, after the general introduction was over. Yes! We enjoyed it very much! There seems to be a lot of interesting people going with us! Said Peter, beaming. Where are you going? He asked. Varanasi. I said, and veered my gaze to where I new the lady of the East was sitting. She was gone! What for? Peter asked. Excuse me?! I said. What for are you going to Varanasi? Said Peter, again. Well, holidaymaking!

Julian and Cecil entered the corridor. Why did she disappear? So, suddenly! Julian felt biased because he liked her looks. Daddy?! Asked Cecil. Yes, my darling! He said. I want to go with David to the Zoo, when we get to the Holy City. Certainly, Cecil! Answered Julian.

Ah, Monsieur! Ah, you gave me a fright! Said Julian. Here are your extra sheets! And something for the tooth for your little sweet! Thank you! Julian felt indebted to this man. I'm not as little as I look! Said Cecil, I'll be 12 this year. I, see! Said the youngish man, apparently expecting an another tip.

IX

Do you know, if the train stops for the night? Asked Julian. Oh, no, Monsieur! We go straight to the Holy City, without any stops! And the train gave an ominous thump. By the way… Carried on Julian. Yes, Monsieur?! Asked the youngish man, again. You wouldn't know the name of that... Well, there is a lady that looks as if…. Oh, Help! Cried a voice down the corridor. They ran to where the voice was coming from. Ah, Monsieur Ferber! What in all monzoons happening! Ah, nothing! Nothing! Just my luggage. It fell down all of a sudden! By the way, you wouldn't know what this is? And he proffered a beautiful hair-brush made of ivory and inlayed with a red stone. Julian's heart leapt as he hadn't seen such a beautiful thing outside galleries. I don't know said the train-guard, apparently uninterested in the value of the find. I found it under my seat! Said Mr. Ferber. Well, never mind. Good night, sirs! Thank you for your assistance! And he closed the door.

Julian was lying in bed. It was close to eleven. He did not know what to expect. All the passengers seemed nice and strangely familiar. He felt as if he was beginning to belong to this part of the world. His mind was constantly turning back to his wife, he left in Calcutta. His daughter Cecil was long asleep. The Miracle of the East was a legend. It was said, that the jewel was so beautiful, you couldn't hold

X

--Page 30-------------
----

your stare at it in a broad daylight. Julian felt he was coming close to things. The jewel was supposed to be buried in a secret chamber in the Holy City of Varanasi. A place, where only the knowledgeable had an entry to. He never dreamt of being so close to something so fabled. Suddenly he heard a noise! Something, or someone was moving above him! The trains bound North carried a lot of passengers, and it was possible that there were people on the roof, as well. Julian fell asleep.

--

The next day was uneventful. Julian got to know Mr. Ferber over the breakfast. He was a jovial man of a broad history both in the West and here. He was a shoe-seller of a family in Venice. An old-school archetype of a tradesman. He had also a history as a correspondent for the Herald-East. A newspaper widely known to the Indian expat families.

Can you show me the hair-brush again? Asked Julian over the coffee. Yes, certainly! I have it in my breast pocket. Julian looked at the intricate object.

Ah, she is not here! Mumbled Mr. Ferber. Who? Asked Julian.

Ah, Jessica, the daughter of the music teacher

Madmun, from the Jaraja family.

Who? Enquired Julian.

XI

Madmun is a name in Calcutta. Jaraja family stretches out to prehistory. They own lands outside the city, and are famed for, well, spawning good musicians.

Julian lost himself in thought. So, the beautiful woman from Dargeeling was actually from a musician's family in Calcutta. Was the man she had such a heated dispute with, at the train station, Madmun?

Suddenly she appeared. She poured herself a coffee and looked strongly confused. Mr. Ferber waved her over. Jessica! I want you to meet Julian, and his charming daughter Cecil.

Glad to meet you! Said Jessica and tried to put herself together. She beamed at Julian. Is something the matter, dear? Asked Mr. Ferber. Oh no! She said, and smiled feebly. Cecil! She said.

Julian is a, well… He travels to The Holy City!

Holidaymaking! Corrected Julian.

The Holy City is a wonderful place for rest! She said, again. Excuse me, uncle! I have not had my breakfast, yet.

Excuse her! Said Mr. Ferber, when Jessica

departed. She is young!

--

It was dinner. Julian poured Cecil the tea. Have those biscuits, Cecil?! Who knows if they have such

XII

34

in the foothills of the Himalayas. What are foothills? Asked Cecil.

They have! Said Peter. Would you like to join us for a small talk. Yes, certainly! Said Julian, and as he was rising, he noticed that Jessica, to Pierre's and Charlotte's consternation, was crying. What has happened? Asked Pierre, leaning his hand against her chair. Tell us, darling! What has happened?! Asked Charlotte.

I'm unhappy!

What are you unhappy for, darling? Asked Mr.

Ferber, coming onto the scene.

Julian came over, and Jessica looked at him as if it were his fault. Come Jessica! I guess, you need to talk to me about something!

And in this way, we got all the tickets to sail to America! Finished Peter. Everyone laughed. So, you say that Yankees aren't such a bad sort, after all! Said Pierre. The door to the dining room opened and Mr. Ferber entered. She's asleep! He said. Good for her! Said Charlotte. I guess, too much emotional stress! Said Peter. You never know! Said Mr. Ferber. Pierre fell silent, as if indulged in thought. I guess we'd better go as well, said Julian, and looked at Cecil. Cecil was already nodding.

--

XIII

--Page 33-------------

Julian woke up and looked at his watch. It was half past one. At first, he didn't know why he was roused from sleep, but then he heard it, again! Something, or someone was moving above him! This time, curiosity surpassed his unease. He crawled out of his bed. Cecil was deep in slumber. He stole into the corridor. Everything was silent. He put on his jacket and silently crouched by Jessica's door. He thought everything was alright, but then he felt a cold draught rising from under the door. He slowly pulled on the door-ball.

Unmade bed and darkness. The window was open. She must have gotten onto the roof. Julian got back to the corridor and tip-toed towards the end of the business-class wagon. There was a small iron ladder, attached to the wall, leading onto the roof. He climbed up and found himself gazing at the stars. The sight for a moment mesmerized his senses.

You have to tell him! Said a voice.

I can't! Answered another, apparently Jessica's. Your uncle should know that we are together!

--

At breakfast, everything seemed in order. Even Jessica was smiling and occasionally glancing at Julian and Cecil. Pierre seemed in a wonderful

XIV

--Page 34-------------
----

humour. He was a poet from France, and had all the looks for her twenty years younger fiancé.

Julian looked out the window. They were having bread and margarine, along with cocoa so good, that almost everyone had two cups. They were nearing Varanasi.

Just two days, said Mr. Ferber. By the way, when we are there, you have to promise to come with your darling daughter to my Mansion. I live just about a kilometre from the Gats. Certainly, we will be delighted! Said Julian. By the way, does the name the Marvel of the East tell you something.

Mr. Ferber looked at Julian, as if he was considered a stupido right from the beginning. But of course, he said, lowering his voice. The Marvel of the East is a legend! It is said that the necklace holds a power to heal the disabled. The tale is as old as Jerusalem. Or, let's say Bombay, and he looked across the aisle, where Charles and his wife Albeda were sitting. As if he did not want them to listen.

Can you tell me where I could find it! Asked Julian Hold your horses, my young friend! Said Mr. Ferber, and he looked angry.

Well, as I said, the Marvel of the East is just a Legend. But the legend says that the necklace should be hidden in the Chamber of Kali. The goddess of death, in the central Varanasi. No one has an entry to that place, except the knowledgeable. Supposedly,

XV

--Page 35--------------
----

the necklace was once in a museum in England, carried there by a stealing archaeologist named Gerry Brownman. The exhibition was surrounded with mysterious accidents, or murders if you please. The necklace was then stolen and returned back to the Chamber.

Tell me Mr. Ferber. How can a healing object be

surrounded with death?! Asked Julian

I don't know, Julian! You seem to want from me more than I can give you. Besides, I'm neither an archaeologist, nor an art seller.

How do I get to the Chamber? Asked Julian,

impatiently.

You have to remember that we live in a modern age! Said Mr. Ferber. An exhibition is a good chance to see precious oddetry. Excuse me! I must speak to Jessica!

Are we going to find the Miracle of the East? Asked Cecil. Maybe! Said Julian. He was again engrossed in the beautiful Miss. Jessica.

--

Julian was in his coupe. He was trying to summarise his thoughts. It was getting dark and Cecil was nowhere to be found. David was probably showing her his collection of stamps. So, it was! The ruby necklace, as the Miracle of the East was

XVI

---Page 36-------------
----

known, had been sheltered in a temple of death. Julian was impatient to get to the Holy City. Only two days of traveling in front of them! He felt as if there was beginning to be a bit of submarine fever palpable among the passengers. Cecil was innocent. But, what about Jessica? Her boyfriend was hiding on the roof, or in the wagons for the poor, in the least. Was he supposed to tell Mr. Ferber? She was expected to be married in Varanasi to a boy of a rich family. Her father's suggestion, apparently. There was no way he could stop this from happening… Suddenly he heard shouts, and a woman's crying! Julian ran into the corridor. Pierre and Charlotte were standing by the door. Mr. Ferber was inside. What has happened?! Asked Julian. I went inside to give her the dinner! Said Charlotte. She did not appear in the dining room. When I entered, she was on her bed, as if sleeping. She constantly mumbled a boy's name! I couldn't wake her up!

Call the doctor! Said Mr. Ferber, as if waking from a daze. Certainly, certainly! Said the train-guard and scurried off for help. Mr. Ferber was apparently angry. The party was waiting for about ten minutes. Then Pierre and Charlotte left. Julian went to the dining room.

He saw the medic Inman only from a glance. He looked eighty and one hundred percent Indian. He was fetched for from the adjacent wagons reserved

XVII

--Page 37-------------
----

for the poor. Then Mr. Ferber appeared. He looked eased off, but tired.
The Jalahar said she will be alright. What is Jalahar, asked Cecil. She
and David were having a toast and butter with cocoa. Jalahar is a Hindi
word for a doctor. Said Mr. Ferber and smiled feebly. Julian began to
understand how little he knew about this foreign land. He said, she most
probably wanted to take away her life! I don't understand it! She has her
wedding fourteen days. Said Mr. Ferber, and poured himself a scotch.

Julian's mind was racing. He saw the man on the roof only for a moment.
What if she does not want to get married?! Said Julian suddenly.
Nonsence! Said Mr. Ferber.

--

The train rolled into the station. Peter and his wife were jubilant.
Pierre descended with the nonchalance of a duke, holding Charlotte's hand
as he went. Cecil was helped by two young Indian guys who were unloading
the luggage. Finally! Breathe the air! Said Pierre. The Holy City is
waiting to be explored! Julian caught his daughter as she jumped out of
the train door. Come! We have to find a hotel. Of some sort… He added,
and looked at Mr. Ferber.

XVIII

--Page 38-------------

Jessica was all smiles. She looked at Julian and, for a while, their eyes met. It seemed that she was coming to terms with her destiny.

As I said, Julian… Said Mr. Ferber, when he was handed his suitcase. There is a plenty of room in my humble abode. Thank you, Mr. Ferber. Said Julian. We will think about it.

I was pondering on what I saw. There was a group of small kids, most probably from the North. They looked funny in their coloured clothes. As if they were coming from far in the Himalayas. I was watching them laugh and lost myself in thought. Cecil woke me from a daze by tugging my sleeve. We were alone.

--

I have to tell you something, Cecil. Said Julian to Cecil in the hotel lobby. They were about two miles from the gats. On the other side of the shore from the Ferber's Mantion. Me and your second Ma. We don't go as well together as I thought. He said, and hoped that she would understand. A man in a suit and a red napkin in his breast pocket came over. Your room is ready, Sir. He announced. We reserved you a number 12.

XIX

---Page 39-------------

Julian and his daughter were having tea in the dining room. Would you care for a biscuit? Said Julian, and pointed to the plate they were sharing. No Daddy, thank you! She was falling asleep in front of her cup.

Julian was examining the map of the city. So

tangled he could barely find his bearings.

A telephone for you! Said a waiter, as he came over. Julian was surprised. He did not expect any calls. He rose and went over to the talking cabin. Hey, that's me! Who?! Ferber! Ah, Mr. Fereber?! How are you! Julian, my niece is not well!

Well, Mr. Ferber. I'm not sure what you mean. She did it, again?

Did what?

Well, I thought she must have eaten something bad! She was alright, and I thought that everything would be just great! Julian never heard him speak like that. You see, she was flicked to the hospital! I'm scared

Julian! He finished.

--

The sun rose like a shining ruby over the horizon. Soon it was so hot, you could barely walk in the open space. Julian and Cecil were sight-seeing.

XX

---Page 40-------------

Come here! Said Cecil, and pointed to a large stoopa next to an ancient house. No! No! Said the woman- guide. They were in a group of five. An oldish, well rounded American, a middle-aged pair from Huston, and two women from France. Apparently, a mother and daughter.

They entered the house. And this… Said the Guide named Janette. Is a house of Madmun, from the Jaraja family. I thought that Madmun was born in Calcutta? Said Julian, perplexed. Oh, no! Said Janette. He was born in the Holy City. He only became famous in Calcutta! Here is his study, and she ushered them into a small, bright room. There was old furniture everywhere. Julian was particularly taken by his library. He had not seen so many volumes of so resounding names for a long time. And here is a small toilette cupboard, which is particularly interesting for… Where is it?! Janette got confused. There is usually a brush which is very rare. She mused. Well, I guess, it must have been taken to the repairs. The oddments on this cupboard are particularly interesting, because they belong to the Madmun's first wife Jiarala. She was taken captive and executed when she was twenty, for refusing to participate in a movement that had its primary goal in eradication of the old knowledge of the holy scriptures. She had been raped and her

XXI

--Page 41-------------
----

naked body was seven days for show on the Kilaka square.

I've never heard of this movement! Said Julian,

surprised.

The movement was called Miata. Janette continued. The Ripe Garden. There is a strong link to Miata and the Miracle of the East. I cannot tell you more, unfortunately, because our time here is finished. I wish you to enjoy your stay, and if you have any further questions, I will be in my office.

Julian spent the afternoon with Cecil in a Café. He wanted to piece this jig-saw himself. Of course that Jessica was Madmun's daughter. But how was she, or the mysterious boyfriend, who so loved the dangerous ride, connected to Miata? The hair-brush was a sign! But of what? Where was the Chamber of Kali? And where was he supposed to find a better accommodation?

Good evening! Said Janette. Suddenly appearing out of nowhere and wearing a shawl and jacket. Ah, good evening, Mrs. Janette. Miss! She

corrected him.

I'd never believe she would appear! She looked so beautiful in the Indian twilight. Her blond hair was falling to her waist. She probably spoke several languages.

XXII

---Page 42-------------

Are you waiting for someone? Oh, no! Said Julian,

unable to look elsewhere. Cecil sighed.

But tell me, are all women so beautiful as you? Said Julian, holding a glass of vine. Don't be silly! Answered Janette. Then, she opened her bag and pulled out a lipstick. Where can I find the Chameber of Kali? Asked Julian.

The Chamber of Kali? Answered Janette. And her voice got a strange tone. As if she were surprised and angry at the same time? Why are you mentioning this place?! She asked.

I want to see it! Said Julian, and raised his glass. See what?! Asked Janette.

The Miracle of the East. Said Julian blandly.

You cannot see it. It was stolen! She answered, this time not hiding her frustration.

Stolen by whom?! Garry Brownman? He said, mockingly:

Oh, no, you silly. Stolen by the army! Two months ago.

Tell me more about it? Said Julian, this time curious. It's been twenty years since the Indian chunta began growing in power. They are constantly on a vigil. Now, they basically rule the city. Suddenly, Julian felt that something was happening. The air grew on strength and then subsided. The sky as if grew darker.

XXIII

--Page 43-------------

Five military men walked among the parasols. The front one spoke. Julian noticed his flags. He was a general.

A pleasant day for a night outside?! He said to

Janette.

It certainly is, Harry! Said Janette, trying to hide

her consternation.

He lowered his head, so that he and Janette met on

a level.

Is this man bothering you?! He asked.

Oh, no! Stammered Janette. He is just a customer

from the office!

Is he an American? Asked the general.

He certainly is! Said Janette, and looked askance on

Julian.

Julian Stanton! From Boston! Said Julian, and thought of his mother in Stanford.

--

Julian woke up. At first, he didn't know where he was. Then, his memory came quickly back. Janette was in the kitchen. Julian went out onto the balcony. He looked across the city. About half a mile off, he could discern the Hotel where Cecil was hopefully still asleep.

XXIV

---Page 44-------------

Janet was preparing the breakfast. Do you like

cheese and onions? Asked Janette.

I love them! Said Julian.

Suddenly, the telephone rang. Janette picked it up!

Ah, daddy! She was speaking…

Yes, but certainly. I know. Hold on.

Janette handed Julian the receiver. My daddy wants

to talk to you!

So, soon! Said Julian.

Hello?

Ah, M-Mr. Ferber! I wouldn't!

Shut up, Julian! We have to find it! Mr. Ferber!

My niece is dying! My daughter will help you. They were in the Hotel, Julian! The chunta is after you! But, what about Cecil?!

I'll take care of that! Be quick! We have no time!

--

There was a hard rapping at the door. Julian and Janette were already on the street. Come, there is a car behind the house!

They sped toward the east. The Chamber opened at nine. I thought you were saying that it was stolen! Shouted Julian over the uproar of the engine. I was

XXV

lying! Said Janette. But Mr. Ferber…! Daddy does not know!

Does not know what?! They got to the sidewalk. Suddenly, there came a shooting. A gunman was on the roof. Another was coming close from the street. Come, said Janette. They entered the Chamber. Janette pulled close the heavy door. There was a large beam used as a padlock. Julian lifted it and stuck it into the sockets. Come! It's underground.

Hold! Hold! Hold! Said the voice of the general. He loomed up from the darkness of the side stairway, leading onto the Citadel. He was holding a gun. In a moment he had Janette in his grip. Don't move American! This won't hurt you!

Julian jumped into the shadows and crawled along

the floor under a flurry of ricocheting bullets. You've betrayed me! Said the general to Janette. I don't know what you are talking about! Said

Janette.

Where is it?! He growled. I don't know! She spat.

They were ascending the stairs to the Citadel. Julian got to his feet. The Citadel was a trap. He looked around.

Come darling! Mumbled the general as he was pulling Janette up the stairway. Jannette was trying to loosen his grip, but he was strong. They got to the

XXVI

top and entered a vast space disappearing into the sepia grays. There was a large bell hanging from the rafters. He shoved her against the floor. Tell me, where did you hide it?! You'll never find out! Cried Janette.

I'll kill you! He said. Do so! She answered.

The general pointed his gun at her. The trigger was moving. Suddenly, something swished in the air. Julian swung on the bell rope out of nowhere and tripped the general as he fell. The gun bounced off. I'll kill you! Said the general as he was rising, and pulled out a knife.

Julian reached under his belt and pulled out a large dagger! It was a fight for life. Suddenly, the general swung his blade at Julian and an ominous boom echoed the space. He fell to the ground, unmoving.

Where is it, Janette?!

Janette was lying on the floor, she was apparently wounded. Julian crouched by her side. He leaned toward her face. Then he noticed it! The Miracle of the East rested on Janette's neck. I love you, Julian. She whispered. Me too! I said .

XXVII

Suddenly a tremendous roar filled the air. I stepped into the bay-window. There was a chopper coming close! Mr. Ferber stood to his word.

Where are we going? Asked Julian. To Bhutan! Said Mr. Ferber. The chopper was gliding through the clouds to avoid radars. The pilot looked at the dashboard. So many controls you have not seen on a rock concert. He was holding a steady height, temperature and the out-flax of gas. Jessica was onboard, as well. Then the pilot spoke. Roger! Roger! He said. And Julian immediately recognized the voice. He was the man of that outstanding English from the train.

Julian thought for a moment. He wanted to piece this jig-saw, again. Then, he asked Mr. Ferber. How did you know that we would be in the Citadel? Mr. Ferber mused. Sixth sense! He said. Cecil was looking out the window. They were flying over the mountains.

The chopper sped like a stallion from an advertisement on drinks. We will soon be down for a descent! Cried the pilot in his Queen's English. Jessica was sick again. It seemed that she was suffering from a disease which only true love could heal and safely eradicate. AARRRGGGGHHHHHHHHHH!

XXVIII

The helicopter crashed into a large patch of trees. The propeller flew about a hundred yards eastward and stuck itself into an enormous jungle-redwood. Is everyone alright! Asked Julian. What has happened?! Asked Janette? We had an accident! Said Julian, and helped them out of the broken up chopper. Wait! Said Julian, and started toying with the door. It was hanging on three hinges instead of four. What is it? Asked Janette. Ah nothing! Answered Julian, engrossed in his work. It's just we were close to fall out before we actually fell! The pilot was dead. He seemed to be passing through a strange sped-up decay, because the only visible remnants of him was his skeleton. Whatever was the case, the Englishman forgot to send out a mayday warning. Probably to their advantage.

We have to carry on on foot. Said Mr. Ferber. Showing strong signs of wear. His chin was sagged and his spectacles were as askance as the Leaning Tower of Pisa. We cannot carry on without supplies! Said Julian. The trip would be too dangerous.

I went around the chopper and wrenched open the back door. It stuck in my hands as I pulled it open. I chucked it to a side. There was a whole bonnet full of food and drink. It seemed that the pilot with

XXIX

Queen's English had a reason to die. It looked as if he were smuggling carrot soup to adjacent countries.

There were no carrots in Bhutan, and soup was valued just as a cheap work hand. I pulled out two bags from under the passenger seats and filled them with stuff so good, it would make a French table for the whole Nigeria. One I gave to Janette, and the other I took myself.

Julian wrenched off the radio-receiver. It stood in his hand like an object of the disaster. He fondled it and put it round his waist. It will work with my Brown shaver! He announced. The party had four members. Just like a hand without a finger. The dick in the cockpit was out of game. Gone. Disappeared! Where now?! said Janette. We have to push through the jungle! Said Dr. Ferber. Cecil seems to be at ease. Let's give her a torch-light! She will lead the party! Said Dr. Ferber. I'm too old for this monkey-game! When it gets dark, she will put it on! It was still light, and Cecil was leading the party. She found a map of Bhutan under the skeleton's hand in the cockpit. The Jungle on the map looked like a wide strip of green.

like this
/////////////////////////////////////

XXX

--Page 50-------------

52

The party progressed through the dense underbrush like an experienced soldier. Jessica was keeping her pace, but she looked tired and way-worn. How long, yet? She would ask. I don't know! Julian would say. Thinking about French tables.

He had in mind a feast for the king of Bhutan, just as the sun slanted westward, and the jungle as if grew dense like the milk of a tired cow. The forest started waking up! Shreeks and creaking were palpable at every corner of the Jungle. A large baboon swished in front of them on a liana, and disappeared into the surrounding shadows. Cecil put on the torch-light. They were on the border of a large rocky strip. A high slope as sheer as a butter cube was looming out into the darkness. Let's stop here! Said. Mr. Ferber. In the yellow light of the torch-light, charged by to alkali batteries, he looked as haggard as a mouse. Good idea! Said Janette. Help me pitch a tent! She said to Cecil and Cecil put the torch-light in between two stones. So that, the beam of the light from the torch-light shone upwards.

They pitched two tents, in all.

One of the tents looked like this

-

/ -

/ -

XXXI

/ -

/ -

....

Mr. Ferber and Jessica were in one tent. Cecil, Janette and Julian in the other. The night progressed. The moon showed behind invisible clouds. The jungle-forest was a tumult of wild sounds. I have to tell Julian what I know! Said Mr. Ferber to Jessica. What do you know? Asked Jessica. Her state was improving with rest. She was eating a grapefruit. That general Harry is following us! How can you know! Asked Jessica with a surprise. I don't know! Answered Mr. Ferber. I feel it!

Julian was playing with the radio receiver he untied from his waist. The gadget was now connected to his Brown electric shaver, and emanated sounds which to Cecil seemed like the humming of bees. Any signal? Asked Jessica. She was preparing a French table.

No signal, so far! Answered Julian. He had in mind a French table for the king of Zambia. He visited Africa when he was in the army as a corporal. He still knew how to survive in a tight. Cecil was examining the map. She was decided upon understanding it before the dawn came. Then, without making any notes of their progress, she fell asleep.

XXXII

--Page 52-------------

They feasted without Cecil. She was way too asleep to eat. They particularly enjoyed the carrot soup. It was, indeed, so delicious that they both, after they finished eating, ran their tongue from side to side of their lips.

Jessica was talking. Tell me no lies! Said Julian after a while. I'm telling no lies! Said Jessica in an injured tone. The Miracle of the East! He said. What's with it? She Asked. It's on your neck! Said Julian, and rubbed his forearm against his mouth. How do you know! Asked Jessica in a surprise. I noticed it, when you were lying on the floor of the Citadel. Said Julian. Oh, gosh! She sighed.

Does daddy know? Asked Julian. Know what? She asked.

That you are a thief! He answered.

I'm no thief! Jessica retorted. Five years ago, when I came to the Holy City, I was a girl without any knowledge. The monastery took care of me! They gave me something which my daddy could not ever afford. What was that? Aked Julian, engrossed in her speech.

A freedom! Said Jessica blandly.

I see! Said Julian, and started undressing himself. You! She said, and laid herself naked into her sleeping bag.

--

XXXIII

--Page 53-------------

The morning came. The sun rose like a red, burning ring. It seemed that they progressed deeper into the jungle than they expected. The breakfast was prepared. How did you sleep, Julian? Asked Mr. Ferber. I had a stone right under my head! He finished.

Julian toyed with the radio receiver. The forest around them sounded ever more lively with animal sounds. We should climb the rock! Said Jessica. I don't think we should! Said Mr. Ferber. He seemed to be gnawed by terrible troubles. Where is Jessica? Asked Julian. Much relieved that he had a partner in a tight situation! I don't know, said Cecil. She went to fetch some wood. She will be right back! Said Mr. Ferber.

I don't think we should climb this rock, either! Said Cecil. The map shows… She never finished the sentence. From the forest, there came, all of a sudden, such a cacophony of shrieking of such magnitude, that the whole party froze with fear. Jessica! Whispered Mr. Ferber. No! Cried Julian, and ran toward the tumult.

He ran to a small space among tall trees. There was a young baboon standing like a boy next to a disturbed pile of twigs. He was a foreboding of what was to come. He looked at Julian with his human

XXXIV

---Page 54-------------

56

eyes, and then disappeared into the under-wood. Janette was gone!

We have to find her! Said Jessica. She was growing on strength. We cannot go so deep into the forest! Said Mr. Ferber. Julian was beside himself. What's wrong, Ferber! Have you lost all sense! As far as I know it's your daughter we are talking about! Cecil was trying to calm his despair.

We cannot go! Said Mr. Ferber

Indeed, you can't! Said the voice of the general Harry.

He stepped out among them, flanked by two soldiers. He was holding a gun. It's not so difficult to find birds fallen out of their nest, when the most stupid one plays with toys he is too young for. And he looked at the radio receiver. Ah, gosh! Said Julian. What are you going to do to us! Kill us! Julian was beside himself.

Where is Janette, daddy! General Harry asked Mr. Ferber. Daddy?! Cried out Cecil. That's why the fear! Said Julian blandly. I woudn't tell you, even if you threatened me with death! Said Mr. Ferber. Run! Cried Mr. Ferber, all of a sudden. The party spread at different ways into the forest. They ran as they could among the trees, eschewing the soldiers

XXXV

---Page 55-------------

57

who were right on their heels. Julian had a problem with a particularly nosy one, who was constantly in his pursuit. He hid himself behind a rock, sheltered by several jungle-pines. The soldier missed him by a few feet, as he ran on. He found Cecil, later on, hiding in an underbrush, and as invisible as a scorpion.

Come Cecil, we have to find Mr. Ferber and Jessica! He said. They cannot be far. They spent the whole afternoon walking through the jungle. An occasional flurry of shooting only ascertained their belief that their two friends were still alive.

Then, with the coming of dark, everything subsided, and the forest started to grow on strength, as the animals began to wake up. We have to be quick! Said Julian to Cecil. We don't want to get stuck in here, when the night comes!

They progressed yet further, into the jungle. Occasionally having to bypass their straight path for the lack of space. How far yet? Asked Cecil, when she could barely walk. Suddenly, Julian stopped. He crouched to the ground and picked up a button. It didn't belong to Jessica, but to Janette! Her disappearance left a cold in his heart. Was she abducted by general Harry?

Look! Cried Cecil, and pointed in between two large palm-leaves. They saw a light glimmer in the distance.

XXXVI

--Page 56-------------

Glad you've found us! Said Mr. Ferber. Jessica was huddled by the fire. She was smiling and gave little attention to them. They were in a large rock outcrop, so deep in the jungle that they lost all hope of ever finding their way back.

Isn't it dangerous to light fires here? Asked Julian,

taking his place next to Jessica.

Oh, no! Said Mr. Ferber. We are sheltered! And in this way, wild animals won't bother us! He finished.

What about the soldiers? Asked Julian.

They have a camp on the other side of the clearing which stretches northwards. Said Mr. Ferber. You seem to know your way around! Said Julian, surprised.

I'm not here for the first time! Said Mr. Ferber. Come! I have to show you something!

They progressed deeper into the sheltered outcrop. All of a sudden, Julian got it. It wasn't just a rock over-hang, but a large cave, leading deep into the rock massive above them.

--

The cave wound itself through the jungle like an underground river. They seemed to walk form miles. Mr. Ferber was holding a resin torch. He looked as if he knew his way.

XXXVII

---Page 57-------------

Where are we going? Asked Cecil. She was

growing uneasy.

Wait! Wait! Said Mr. Ferber. He was trembling with excitement, and Julian for a while thought, that this old man was not all there, after all.

There! Mr. Ferber pointed. Where? Asked Jessica.

They pushed through a narrow corridor and suddenly found themselves in the open. A vast city of apes laid itself prostrate to their sight. The ruins seemed to stretch themselves for miles. Thousands of apes inhabited the ruinous magnanimity.

They walked half dazzled among the animal citizens. Julian thought that he was dreaming. Jessica was smiling and proudly walked behind Mr. Ferber. Cecil was last, chasing away curious nosers with a stick. Julian for a while thought that Jessica must have been under the influence of some hallucinogen. Or maybe they all were. He felt himself pulled to Mr. Ferber, and the more to Jessica. His imagination raced with wild dreams of Jessica and Janette as his lovers. Suddenly he fell on his knees with exhaustion.

There! Mr. Ferber raised his hand.

They were standing in front of Hanuman. The king

of apes.

-

XXXVIII

--Page 58-------------

- - -

---------------------- - - - - - -

O, Hanuman! Cried theatrically Mr. Ferber. The whole party was as if in a trance. Janette was sitting next to the king, on a throne made of an intricate weaving of palm roots. She was looking in front of her, apparently unaware of the friend visitors. The marvel of the East rested on her neck. Julian looked at it. He desired the ruby necklace more then his life, but he couldn't hold his eyes at it for more that a short while.

He became angry and tried to go to Janette with outstretched hands, as if he wanted to take it! Two large apes barred his way with spears. He fell on his knees and started begging for his life.

I've brought you the blame ones! O, Hanuman!

Spoke Mr. Ferber.

What have they done? Asked Hanuman.

They trespassed on your land, and almost thwarted the goal! Said Mr. Ferber. He spoke like a man without a reason, and saliva was visible in the corner of his mouth.

XXXIX

---Page 59-------------

They will be executed! Said Hanuman.

Oh, no! Said Janette, suddenly turning her eyes at

her capturer. They belong to me!

How, come! Said Hanuman, turning to her.

I want them as a remembrance to my former life!

She said.

I don't think I can allow this! Said, Hanuman. He

seemed to be mesmerized by her beauty.

Come to me, darling! Said Janette.

Hanuman, made a few ape-like steps and was at her, stroking her hair.

Suddenly, Julian noticed it! The precious hair-brush rested in Hanuman's hand. All his memory at once flooded back. No! He cried. But Janette was quicker. She inconspicuously pulled out Hanuman's dagger, as long as a palm-shoot. And when Hanuman turned to Julian, he fell dead on the spot.

The whole party woke up from a sleep. They were in an expensive hotel on the border of Bhutan and India. Close to home. Janette was drinking a coffee. Julian was writing to Singapore. He wanted to tell his second wife that he wouldn't be coming back for the holidays. Mr. Ferber was playing a board-game with Cecil and Jessica.

Here you go, again!

XL

--Page 60-------------

XLI

--Page 61-------------

„There is a legend which shakes the fundamentals of the far East. Will Julian and his daughter Cecil reach the Holy City to search for a jewel which

might save their lives. A story leads to a story. Allen Petterson is a Classic?"

XLII

---Page 62-------------

Gill Bonmouth

Benjamin Schmidt alias Allen Petterson

was born in 1983. He studied at the FF, UK, Prague. His profuse traveling brought him to writing. He wrote On The Count of Three (Až napočítám do tří), and a collection of Haiku. He is a song-writer and a musician, and currently resides in Prague.

XLIII

---Page 63-------------

Jessica looks distraught. Her wedding is coming close. Julian seems to abide by his place, as he searches for the Marvel of the East.

Gill

Bonmouth

XLIV

---Page 64-------------

66

An amazing story for all ages..

Lui Bridle

Gallerymarvels.com

XLV

---Page 65-------------

.

XLVI

--Page 66-------------

XLVII

--Page 67-------------

XLVIII

--Page 68-------------

XLIX

---Page 69-------------

L

---Page 70-------------

Gallerymar vels.com ¨

LI

--Page 71-------------
----

with all rights reserved.. Allen Petterson

Publishing

LII

--Page 72-------------

Index

1.

F OCLÓIR G AEILGE – B ÉARLA

O BAIR B HAILE – HOMEWORK , PAGE 50.

2.

I RISH P ARLAMENT , I RELAND ' S TAOISEACH , E NDA K ENNY ,

6.

J ON C OHEN , A LMOST S HIMPANZEE , USERPTS FROM THE LIFE OF J ANE G ODALL ,

D IFFERENT PERSPECTIVES ON SHIMPANZEES AND FARMS . P AGE 55, 62

7.

T HIS IS A LIST OF

WOMEN

T EACHTAÍ D ÁLA (TD S).

C AROL J ANKINS , N OVOTNY , D NA TESTING ON TRIBES

F EMALELEGISLATORS , SENIOR OFFICIALS AND

MANAGERS

(% OF TOTAL) .... I RELAND . 8. 22. I SLE OF M AN .

F AMOUS

OF P APUA . P AGE 120,165.

S HABDAKOSH , H INDI DICTIONARY , J ALAHAR , P URER ,

THE ONE WHO CHASE AWAY

A N EVI SPIRIT . P AGE 120.

NAME IN I RISH POLITICS , 72.

3.

A USTRALIAN ERASMS , PAGE

8.

M IATA , M UD BEATLE , S HABDAKOSH , H IDI D ICTIONARY , TYPE OF WESTERN CAR MAKE ,

J ALOP , MEXICAN SPANISH OPIATE MUSHROOM , ALSO

KNOWN FOR

I TS HEALING PROPERTIES , PAGE 1.

4.

N OTEBOOKS FROM P APUA N EW G UINEA , N OVOTNY ,

PROGRESS IN

T ECHNOLOGY AND MINES , AIRPORTS AND CINEMA , PAGE 30, 56,120.

5.

S TEPHAN M ITTEN , A FTER THE I CE , PUBLISHED THROUGH P HEONIX . L AST GLACIAL

M AXIMUMS , SO CALLED LGM S . P AGE 120, 135.

11. A USTERE LIFE , S RI RAMANA M AHARISHI ,

L IFE .. P AGE 25,126,175

12.S RI N ISARGADATA M AHARADZ , TRANSLATIONS ,

12. B ENJAMIN S CHMIDT , P AGE 25,75,125, 13. I NDO EUROPIAN STYLES OF MUSIC , P SYTRANS ,2005, LEGALITIES PARTIES IN I NDIA ,P AGE 25,75.

14. P ANATIPATA SUTA IN P ALI , ALSO V ESAK SPRING

B UDDHIST CELEBRATION IN EUROPE P AGE 25,125.

15.P INDAPAL , MORNING ALMS , RICE AND FISH IN

P AGE 25, 76,120.

9.

S AURIEL PARK K LEINVELKA , T HE F RANZ G RUSS D INOSAUR G ARDEN

P ERSON WHO BULIT IT WITH BRICKS AND LIVED NEARBY , 1978, P AGE 126,135.

10.

O RIENT E XPRESS , TRAIN CONNECTIONS FROM GERMANY , TO E AST .

A LSO TRANS - ORIENTAL .. P AGE 75,56.

16.

I RELAND ,D UBLIN , G ORILLAS , R OBERTSON F OUNDATION , AWSPICES , P AGE 75,125

17.T RES BIEN , VERY NICE , F RENCH D ICTIONARY .

18. H ARP , C LARINET ,F LUTE , V IOLIN , H ARMONIC , S ITAR ,

G UITAR , R AVI S HANKAR S ITAR , T ARA , SHRUTI - BOX .

M USICAL INSTRUMENTS OF E UROPE AND I NDIA . P AGE

THE BOWL ,

T HERAVADA B UDDHIST TRADITION .

125,350

---Page 73-------------
--------------------------------------- .

11. A LPINE MAN , AN ALMOST N EANDERTHALI FOUND IN A LPS ,

P RESERVED BODY IN ICE . B ATCH OF ARROWS . M ITTEN , P AGE 120,235

19. P EAT B OGS , M ITTEN , BURRIAL RITES , P AGE 25,256,374

20. L OGICS AND IMPORTANCE OF UNDERSTANDING SOFT TISSUE ,

M ARTIN B RASIER , P AGE , 254,275. P RESERVATION .

--Page 74-------------

Tended Stones - Cote D ' Ivore Blueprints

Geza Teleki , 77, died on January 6, 2014 at his home in Budapest, Hungary, International Primate Protection League founder Shirley McGreal ...

Dame Jane Goodall DBE , born in London on 3 April 1934, is

a zoologist . [1]

Goodall is a British primatologist , ethologist , anthropologist , and UN Messenger of Peace . She is the world's foremost expert

on chimpanzees . Goodall is best known for her 45-year study of social and family interactions of wild chimpanzees in Gombe

Stream National Park , Tanzania . [2] She is the founder of the Jane

Goodall Institute and has worked extensively on conservation and animal welfare issues.

William McGuire " Bill " Bryson OBE HonFRS (/ˈbraɪsə n/ ; born

December 8, 1951) is a best-selling Anglo-American author of books on travel, the English language, science, and other non-fiction topics.

---Page 75-------------

Hindi, French and Chinesse Dictionary: Mausam - weather

Ye sundar mausam he. - The weather is nice. Zindagee - Life Cini - Sugar

Ek,do,tin - one,two,tree jutli - banana peel capati - bread zai tien-good by (chinesse) Bjor - Beer

Je ha - Yes (aso in Chinesse) Je nahi - No (also in Chinesse)

larkee - girl maro mat ko chuo - tighten a horse French: eglise - Church

Come sa - how are you ekreer - write je bien - I'm fine lui - read

regarde - watch out z aime la arbour- I like the trees ju - yellow
regarde la eglise - look at the church se pa possible - that's possible

--Page 76-------------

I Re ach for m y d i ary`s End In Ro m e a n d...

up s..

each

End

up

Pr a y ers

Pray

When you`ve reached the utmost of you physicality,

to alter the course of all events.

When you`ve siduced the Devil to your cause,

and in profundity with angels dance.

When prayers no longer suffice, and all old ends, and all new rise..

When your voice coarse through the desert borne, sings a song of old vows
sworn.

When through the fog come seagulls cries,

RISE

RI

--

and something so gentle within you dies.

--Page 77-------------

Dedicated to my pa rents who do es not s pea k so this prolificaly
language .

this long

B.S.

--Page 78-------------

Foreword

In November 2012, I went to Italy to visit my Guru. I had not seen him for 10 years. We hardly talked, but then we sat a lot. He surprised me how good he looked, and I

surprised him in what bad state I was. We got up at 3 and ate once a day.

The rhythm of life is different in that mountain range... at once

once

...I stayed for 7 days

--Page 79-------------

Mo r n ing

I rise at three,

and its s t i l l dark.

I f o rget thee,

In this h i l l y Mark..

l l

--

Br eakfast

Eight hours till ill breakfast,

Some t h I n g t o forg

et..

--

a pi ne-cone

I`m where the pine-trees chime, I`m w here the words do rh y me.

--

W ork

I paint, paint

I p aint.

--

I w as a painter, then..

I`m Da Vin ci,

if you al l o w me..

--

1

z

A S troll

trall

I roam to the Forest, and behind a cone, I take a b

reath ,

amd walk back Home.

--

Of your Dre am..

I don`t don`t drive,

But I walk fast.

I eat little,

But then I savour food;

I`m the man of your Dream , So, I hope that we will sui t.

--

2

--Page 81-------------

Fo o d

I eat Bread ,

And it feels b Bread

e

t t

er the n sex .

--

Letter

Each word I choose,

with the the Tender ness -

The rustle of that leaf outside.

--

Celibac y

slouched

Nine mo n ths untouched, Nine mo n t h s touched ..

--

Se x

If I had s ex ex

I would have had it -

Spread honey on that brea..

d. .

--

3

--Page 82-------------

Wisdom

Once, I heard a man talk wise;

e..

s e..

i

Now, I see the Sun gor

--

Floor

Did I scrub the flo or in s leep, Or, Am I s leep ing now when scrubbing
it ..

--

Drifts

I watch the drifts up on my blanket; The h ea t I n g `s of f this
winter-time.

--

4

floor

dri

---Page 83-------------

Snow

I remember sho velling snow, but I don`t remember an

y..

--

Land

This Land hides a lot of beauty,

This land, too.. lot

--

Hills

The h ills are like b reast s o f so me - pr I m eval b east.

--

No tes s u r around me

Forgot

I forgot I`m here,

I`m D ea f t o this place..

Deaf

--

5

a

---Page 84--------------

Sho p ping

I go sho pping,

t e n thou s a nd burning f l a m e s..

fla,

--

Rome

All ro a d s lead to roam;

They s a roads y,

--

Priest

In t h ese s treet s I`m a p r I est..

--

6

---Page 85-------------

A isle

I seek a mong aisle s

monk

--

Gu r u

I left my Guru in a state of c o nf usion

--

Su n set

Lau gh but not l e ast We l come!

--

Siege

T h is for t r e ss, I can`t defend..,

I

--

He art

To call you b r ave, would be, My weak n e s s.

--

7

--Page 86-------------
----

min

min

Mi s un derstanding

He does not see a snail;

an engry monk snail

--

Tradition

Monastery, I left lie a m a d Man

--

A m on k

Drying m y und erpants;

s

to h is s u r p rise

sur

--

H om eco ming

The re i s noth in g like that..,

B.en

Ben ja

ja

Schm Idt.

Schmidt

8

--Page 87-------------

Revitalisation of a House (Classical French Manor) Tiles - Benjamin
Schmidt

---Page 88-------------

Revitalisation of a House (Classical French Manor) Tiles – Benjamin Schmidt

Twopeople (All

rights Reserved). Roof column and frescoes (doors) facade, adjacent house, palaeontological centre.

--Page 89-------------

Excavations in a Czech subarb. New shopping malls and centres arise.
Inhabited parts of subarb. New economical houses. Housing estates.
Excavations dug up in squares. An archeo/paleontological excavations , a
collegue serches for his tool. Subarb in Italy.

---Page 90-------------

The Beach with the Tended Stones

--Page 91-------------

Julian in a car, racing down with Brun to save Sharlotte.

--Page 92-------------

The story continues..

I think that it would be possible to fill the whole book with pictures
and leave partly for adults. Also the possibility to have those pictures
at the end for children to have a look.

```
------------------------------------------------------Page 93-------------
----------------------------------------
```

-1
--Page 94-------------

--Page 101------------

---Page 114------------

---Page 120------------

---Page 123------------

Last page of the book...

There was not too much I could do with the yogurt. I was staring into the fire and imagining all the past memories from my travels. The cake was baking in the stove. Another one, again. After a long spell of time. I couldn't believe that Cecil came for the holiday to stay with us again for a couple of months in our Shattau Mondo.

The maid came in and told me that the supper was being served by Sharlotte. Oh, I didn't know that we had a special Occasion, I thought for myself. Sharlotte prepared something I understood. Sousages with some strange mushed porrage, or something that reminded me of foreign food from far countries.

Is it spiced with some strange flavour? I asked Sharlotte. Oh, yes Julian. It's a recepie I got from the work last year. Interesting, I said and bit into the sousage. I was cutting the porrage with a fork and I understood the logic of the Stereotipical Landscapes. I understood that Ajhan from the Italian monastery was like the carrot and the sousage was like the Tended Stone. Sharlotte grew uneasy about the supper. And She asked me if I was

--Page 124------------
----

alright. I wasn't sure and progressed through my meal. I was intent on the carrot, that Ajhan would write me a letter that he received the socks, or something. I understood that even him must play occassionaly on the sand to look at sky. I would never realise that to cut the sousage would be so difficult, when you ponder something as hard as a rock.

How far are you with the research, I asked Sharlotte. And I fidgeted with

the fork in the porrage to let it cool down little bit.

Sharlotte was in a good humour and she began to expound on certain gnominy abot the logic of her work and her future progress in career. I understood, I faked the sudden rush of needing to wash a couple of plates and slipped the sousage under my sleeve. I went over to the sink and tried to show how I could dry my hands above the soucepan when the sousage slipped on the floor and shattered itself.

Tough isn't it. Said Sharlotte. What my work, I asked. And washed my hands in the sink. Cecil rushed in and she was holding something I've never seen before. A dragon box from dynasty Ipal Caqua, from Jipirada. Oh, you scare me Cecil. Why don't you play with better things then my old shoe box. Oh, what is it?! Asked Cecil and looked at the floor. The stone shattered in a way I comprehended. I slipped and broke the light bulp with my hand and we remained in utter darkness. The telephone rang. It was the Chief of the tribe from Kipaka (from the Bhutan boundary where I got lost).

There you go again, said Cecil.

---Page 125------------

On the Count of

Three

Prognoses for richness, growth in the community and the society.

PH.D. External studies

2007-2018 2. part of Ph.D. Work coprolaly, problems with speech maladies from uneducation..

Through my external studies I understood a lot from the differentiation between life and the prognoses for obtaining oveview on logical aspects of livinghood, further prognoses on life among monks, Budhist monks and prognoses for life among tribes of African aboriginal community and Papua New Gunea, and Madagascar. My Ph.D work contains my published book in New York and Canada. I further am the author of a collection of Haiku (userpt), and a book of poetry. I strive for publishing a complete science at oxford , or America for further prognoses in art, literature and science for progeny.

125 pages the whole text..

---Page 126------------

On the Count of

Three

Benj a mi n Al anSchmi d t

The path wound itself forth like a thread winds itself out of an old
woman's hand. The tumult of the forest was a mellow hum. The mushrooms
were basking in the late noon sun.

What's that, said a voice. And the other replied, I don't know. Soon, you
could hear the thumping of hooves against the hardened dust. Hooves of a
steed so swift, that the distance could not be guessed. Where to the
Mildred's Chaishleen? Asked a figure just come on a wind, towering like a
smouldering pyre.Who drives you in thy maddery? Whispered the twisted
mouth of a pagan. I'm come from far beyond the Brihin's gaze, and trevel
to my destinaton's end. Advice me,O pagan, lest thy here stroll be the
last to take!

You are on the right track, aya. The Mildred's Chaishleen lies yonder,
behind the birch grove. You'll see the black stone rising high. Than,
just twenty yards and a thousand steps. The devil must be driving you to
enter that Chaishleen. Rumours are the place is haunted by misery.

The devil be with you! Cried the voice of the figure, and, the way he
came, he disappeared. Just as the

cock kreed an afternoon. Driven by some foreign fury. I

leaned over where Lucy was lying. She looked so

despicable in her injury, that I thought she may not survive the day. She
needs immediate hospitalisation remarked Sergeant

Logical stipulations throu linguistical bariers with prognoses to
Pali,Hindi, and Sanscrit.

The clouds are gathering, just for a rain. Where is Penebrall, my dear?
He was supposed to be here at two. I think he might have got stuck
somewhere. Answered Penelope. He likes sticking around places, after all.
Particularly where there are clouds. You know how obsessed he is with all
this new-age commotion about predicting the weather. He may, as well, be
on our roof, playing with his gadgets. Here I'm, Harold! Boomed a voice,
and Penebrall appeared from behind an invisible door, carrying an

2

Curriculum Vitae

Benjamin Schmidt Hošťálkova 61 Prague 6, 169 00

+420 608268129

125 pages the whole text..

--Page 127------------

131

On the Count of

Three

Education : 1990-2003

2003-2004

2007- 2009

Private College of Art and Design, Prague

Old graphic techniques (dry needle, copper plate, lithography, and all kinds of etchings). Mastering computer programmes such as; Adobe Photoshop, Adobe Illustrator, 3D Studio Max.

One year course of English with American tutors -The Language House,

Prague; finished with a diploma.

Study at The Faculty of Philosophy, Charles` University, Prague.

English&American Studies (left for family reasons in 3th semester).

English, Irish Gaelige, Hindi, The theory of Irish and English literature.

Work experience :

Archaeological Art asistant -Archaia, Prague
  (2003,2008)

Working part-time as an archaeological art asistant at the National Museum, Prague. Participating on archaeological excavations (learning archaeological terminology, geology).

Secondary school teacher
  (2006-2007)

Two years as a secondary school teacher of English. ZŠ- Petřiny-sever.

Freelance translator

Currently working as a freelance translator. Translating architecture, art, environmental sciences.

A former Editor for Amnesty International, ČR. Translated venerable life of Arunacala

125 pages the whole text..

--Page 128------------

On the Count of

Three

Underline information :

One year course of Irish Gaelige and Hindi languages. at FF, UK, Prague.

The field of Art:
 my private research...

--   "The
Theory of attitude towards tools of Art" "An Attitude, as a Means for
Self-Regeneration"

Several years private lector and tuitor, lunetic asylum art therapy
theory and education theory. Sewing, knitting, jewl making and pottery,
theory of attitude towards education and remedy. Concerts. Zizkov 2.
Dobra Trafika, Prague 1 Several times, Ireland France. Tin Whistle -
Traditional Irish whistle + dance, dagda

Guitar School theory education, jazz improvisation, blues, theory and
practice attitudes towards practice and theory, notes theory

Indian Sitar - Tablatures, notes, theory of attitude / Jiri Dohnal Bells,
Irish, tibetian, tibetian bowls, triangle, bazuki, Piano -

Also. Photography

Freelance translator

Dosažené vzdělání:

1999 - 2003 Střední umělecká škola Designu, Praha. Staré grafické
techniky (suchá jehla, mědiryt, litografie, aquatinta, lepty).

Počítačové programy: Photoshop, Illustrator, Maya, 3D Studio Max.

Kompozice písma, Typografie.

2003- 2004 Roční studium Angličtiny, The Language House, Praha.

Američtí lektoři -Zakončeno diplomem.

2007 - 2009 Studium na Filosofické Fakultě, Karlova Univerzita, Praha.
Obor: Anglistika-Amerikanistika. (studium Angličtiny, Irštiny, Hindštiny
a teorie anglické a irské literatury a dramatu).

Studium ukončeno z rodinných důvodů po 5. semestru.

Zaměstnání:

Archeologický asistent

125 pages the whole text..

(2004)

On the Count of

Three

Společnost archeologů ARCHAIA

Archeologické výkopy pro dnešní Paladium (Rudolf II.), Veleslavín

Dale. Centrum2/5, germanska kultura, Podbaba,2/5 mes(Paleolit). Zlicin 5 mes geologicke anomalie

Překladatel
(2005- a současnost)

Překlady environmentálních textů - archeologie, geologie, architektura, umění. spolupráce s australskými a kanadskými programátory na tvorbě webových aplikací.

Učitel Angličtiny druhého stupně ZŠ Petřiny-Sever (2006-2007)

1. Výuka Angličtiny podle kurikula.

2. Každodenní příprava na hodiny (1 a 1/2 roku zaměstnaný na nadůvazek).

Datum narození: 19.12.1983

Od roku 2011 se zabívám teorií přístupu k umění.

Currently working as a freelance translator. Translating architecture, art, environmental sciences.

A former Editor for Amnesty International, ČR.

Underline information :

One year course of Irish Gaelige and Hindi languages. at FF, UK, Prague.

The field of Art: my private research...

--- "The Theory of attitude towards tools of Art" " An Attitude, as a Means for Self-Regeneration"

G a l l e r y m a r v e l s . c o m

Several courses of figure and act drawing

Exhibitions: 2,5 years

Born: 19.12.1983

Vystavoval jsem v Divadle komedie jeste za Lucie Vondrackove , pote kousek od B etlemske kaple , Klub Kastan Praha 6 (2005), Majk L ' Atmosphere Praha 6 (2013). Chystam nyni vetsi vystavu s koncertem v Dobre Trafice na Ujezde . PBN Bohnice , exhibition 5 . Kolona

A possible exhibition in Paris / Close to the Jewish Quarter / 2018 - a planed exhibition in Prague Photo gallery Ujezd - (closely connected or linked to Czech Press Photo}

125 pages the whole text..

---Page 130------------

On the Count of

Three

Postcards

A planed exhibition in Gallery Miro, and Rudolfinum/ Postcards

(Visited Sorbonna, FF, Uk, Maggee university Dery, and several other schools in Co. Cork, and Paris. Luvre, Sagrada Familia, Pizza, Giant's Causeway, Eifel Tower).

Znam se osobne s Vladimirem Mertou, Vladislavem Matouskem, Oldrichem Janotou a Vohtechem a Irenou Havlovimi. Jsem dobrym prateli s Jirim Dohnalem, Petrem Korbelarem, Ondrejem Smejkalem, ci Radanou Lancovou.

Snazim se propagovat umeni , ktere ma co rici zkusenemu vytvarniku , i laiku , metodikovy , I exuberantnimu hromotlukovi . Jsem byvaly pedagog , a zabivam se vlastnim dogmatem nastroje jako prostredku k seberealizaci .

Art therapy / Several Lunetic asylums, Ireland, France, Czech Republic.- Sewing, on a sewing machine Knitting,

Theory of a lunetic asylum art therapy, music therapy, lectures on education, elocuation courses theory, and practice. Painting, ceramics, pottery, candle making, broche and precious stone ornaments making, jewl making, artificial beads explicated science.

Lived and visited /- 5 times Italy 2 Times Ireland, France and Spain 3 Times Germany and 1 Denmark

- Scout camps, meditation centres in several countries, and meditation trips, knows several important monks. Bhante Vimala, Bhante Dhammadeepa, Asin Ottama, Santacittarama Monastery. Bhante Vimala / Founder of the meditation centre Lotus / Prague 2001.

Lives partly abroad.

Ve své tvorbě se soustředím na:

(Přístupy k výrazovým prostředkům, Umění, přístup a seberegenerace).

Born: 19.12.1983

On The Count of

125 pages the whole text..

On the Count of

Three

Three

old navigation triangle. Just as I thought, cried Penelope. The Dinner is
soon to be served!

--

The clock beat just four and the dining room started filling with people.
Duke Alabasar had come straight from the Far East. He procured there a
rare disease which forced him to blink. With his thick obstacles, his
eyes were like that of a frightened mole. No doctor in the West was able
to tackle down his problem. Doctor Forlore came, wielding a white cane.
He was an expert on tropical fauna. He demanded respect in circles which
Penelope might have only dreamt about. Then Julian, his wife Groine, and
his daughter Cecile along with his family friend, an alder man, Gruptni,
and his son Jekery. They were from Calcutta. Arrived just seven days ago
to have a look at the city. Holidaymakers. As Julian whispered to his
wife, when they were waiting on the platform.Gees were served, then
lobster on garlic, and the best wine that could be fetched in this
season. The grapes were long ripe, and leaves were putting up hues of
Bertold Brecht. The discussion sifted between giraffes of Africa and
Doctor Forlore's knowledge of nonexistent plants. Then, Gruptni was poked
to brag about his country, and everyone clapped when Jekery recited an
old prayer in a language that no-one understood. It comes from Verepara,
an old city north of Varanasi, our holiest city. Varepara lies about a
hundred kilometres north, at the gates of the Himalayas. I heard that
there are people living on the slopes of the Big Mountains, who gave
their life utterly to self-denial, healthy food, and exercise which, if
I'm not mistaken, is called Pranajama. Said Penebral. 3

On The Count of Three

That's true, answered Gruptni. Although, pranajama is just one of the
many disciplines of

125 pages the whole text..

On the Count of

Three

Joga. People in the North of India are very spiritual, if you don't mind me using the word. Why should we mind? Ask Harold. It's because you here are mostly scientists. In our country, we don't talk about god, when we talk science. If you know what I mean. I know exactly what you mean said Doctor Forlore and stroked his beard. I don't. Said Julian, and his wife Groine gave him a prick and stifled a laugh. You are all too serious for such a feast. Said Alabasar. And besides, how often does it chance to you to be hosted by such a beautiful and lovely creature. He added, and raised his glass to accost Penelope. We gathered here to eat together, and to be merry, he continued. But don't forget that this is not just a start, but exactly a start.

A start of an adventure to be continued. I want to see you all in the study at seven. We'll discuss our journey. Doctor Forlore was standing by the window. Julian was by his side, asking questions. I mean, where does it lead all this scientific progress that Penebrall is so concerned in. Why don't we just live with monkeys on trees like we used to before we decided to live modern. You are speaking my thoughts, said Doctor Forlore. But you must remember that we are part

of this society, and therefore we should abide by its rules. Come, it's seven. We are supposed to be in the study.

They entered, and Julian let the door click in. Penebrall was just expounding on one of his

gadgets, used to predict rain from storm clouds. Gruptni seemed thralled by the idea that, if he chanced to be shipwrecked on a deserted island, he would be able, out of his palm-tree hut, to predict a coming storm. And, therefore, be able to fortify his abode against such an inclemency, before it happened. Penelope began to be inpatient and in a while said. Penebrall is a knowledgeable person, but he might as well begin to be kind and disclose us why we all met here. I'm on a pine-cone with excitement, said Harold. The group seems ready to listen. Said Doctor Forlore, and seated himself on a chair. Groine crossed the room to embrace Julian and Penebrall began to talk.

We are, people, like a body in the intricate system of the human build. Let us just imagine that the skeleton, I mentioned, is the world. And we, as people, abide by its rules. We have to acknowledge that we are human, because we think that way. But what if I told you that, sometimes, our thoughts can be curtailed by what we see.

What do you mean, Penebrall? Asked Harold. Aren't the instincts just what drives us

forward? Exactly, Harold. But that's just what I'm talking about. As the representatives of the human species, we should be concerned more about what eludes our understanding then what we understand. Here, and Penebrall reached under the table and pulled out a map. Is a place where we have to go to understand our pre-thoughts. Our origin lies on this island. And he pointed to a criss-crossed spot on the Map.

Where is it? Aked Julian. It is in Indonesia. South from India. Answered
Penebrall. And how are we

125 pages the whole text..

---Page 133------------

On the Count of

Three

going to get there? Asked Doctor Forlore. Do you know how long it takes a
transatlantic ship to sail to Calcutta ? Said Doctor Forlore. Are we
going to Culcutta? asked Gruptni. I do, said Penebrall. Exactly two
months and fourteen days. How from there? Asked Julian. Then, we will
fly! Said Penebrall.

Fly?! By all Dervish, said Doctor Forlore. Who heard about that! I did,
said Julian. I saw it in France, last year. They heat up air in a
tremendous balloon. I saw it carrying three people. Don't forget that
progress in unstoppable! Said Penebrall.

--

4

125 pages the whole text..

--Page 134------------

On the Count of

Three

On The Count of

Three

I Re ach for m y d i ary`s End In Ro m e a n d...

125 pages the whole text..

--Page 135------------

On the Count of

Three

up s..

Prayers

When you`ve reached the utmost of you physicality, to alter the course of all events.

When you`ve siduced the Devil to your cause, and in profundity with angels dance. When prayers no longer suffice, and all old ends, and all new rise..

When your voice coarse through the desert borne,

125 pages the whole text..

--Page 136------------

On the Count of

Three

sings a song of old vows sworn.

When through the fog come seagulls cries, and something so gentle within you dies. --

Dedicated to my pa rents who do es not s pea k thi s l a nguage . B.S.

Foreword

In November 2012, I went to Italy to visit my Guru. I had not seen him for 10 years. We hardly talked, but then we sat a lot. He surprised me how good he looked, and I surprised him in what bad state I was. We got up at 3 and ate once a day. The rhythm of life is different in that mountain range... ...I stayed for 7 days Morning

I rise at three,

and its s t i l l dark. I f o rget thee,

In this h i l l y Mark.. --

Breakfast

Eight hours till breakfast, Some t h I n g t o forg et..

Mist lay on the sea as they were waiting for the deck to open. Julian's watch were showing five minutes to eight in the morning. It was drawing on the end of fall, yet the breeze was pleasantly mellow, and the seagulls were crying some merry tune.

125 pages the whole text..

--Page 137------------

On the Count of

Three

A sergeant came and unlocked the stairs that lead onboard. The party was complete. All nine members were ascending on the deck. How long are we going to be away? Asked Cecile, holding her father's hand. I don't know said Julian. Don't ask questions that no-one can answer. We are holidaymakers. And Julian gave a sidelong glance at Gruptni. Jakery smiled at Cecil, and then hurriedly followed his father. Come. Said Harold to Penelope. It will soon be time to set sail my dear.

Julian was gazing at the sea. It will soon not be a sea. Said Groine to Julian, and leaned against his side. I know. Said Julian. I'm a bit worried! Said Julian, and leaned his head to where Gruptni and Jackery were sharing an ice-cream. Don't be, said Groine. Young Jackery is not a child anymore. He is. Said Julian. It may be that he will not be. But he is at the present.

Look, cried Cecile with excitement. They are pulling up the anchors. Wave. Cried Julian. We are sailing! 5

On The Count of Three

--

The ship sailed out of the harbour. The huge fog-horns blew and the sun showed up behind the morning sky, as they flew out beyond the buoys. People were merry and men were putting off their hats while ladies stayed with children onboard to watch with binoculars the disappearing shore. Come. Said Groine to Julian still lost in the waves. They are waiting under the deck.

Besides, we have a room 204. Which makes us next-door neighbours with Gruptni and Jackery, if you don't know. What a surprise! Said Julian with a smile, and offered Groine his arm. Seems like this trip will be adventurous, after all.

--

Put the pillow here, said Groine. And sleep with your feet on this side. As you say, general. Said Julian. They stopped. Did you here? Hear what? This… Ah, You scare me once again, Julian, and you are a man overboard. Knok, knok.

125 pages the whole text..

--Page 138------------

On the Count of

Three

Ah, Jackery! What brings you to our small abode? Mrs. Groine, look what I
found! And Jackery reached under his vest. Ahhh! Yelled Groine. What's
that! Cried Julian, and snatched the dagger from Jackery's hand. 6

On The Count of Three

I-I f-found it under my bed! I thought I could keep it. Did you…
Whispered Julian, with his eye on the blade, close to light. It's a
harakiri dagger, as far as I know. You know-it-all, said Groine, and took
the blade from Julian's hand and handed it to Jackery. Thank you Mrs G-.
For you, just aunt Groine. Thank you, aunt Groine! You are welcome. You
are making too many friends, Honey. Said Julian. Don't be that way! He's
just a child. Retorted Groine.

Onboard you were saying something else. Said Julian, and stretched
himself on the bed. Say what may! Said Groine, and laid herself on her
yet unprepared bed. Exactly! Said Julian.

--

 Film competition I.

The film is based on an idea of education, elation and jazz or R&B
production. I'm singing west Virginia

pride in a paradigm - those are mormons who travel to Africa to educate
and spread belief.

I'm coprolitic in my expression / Book 1. as well, - and I claim that I'm
trying to use swearwords and demuring all-recital language- to solve
racial problems.

 . .

Famil
 . . .
 Enfant

125 pages the whole text..

--Page 139------------

On the Count of

Three

- The film is devided into ½, 2/3 segments (famil, enfant / family,
children.. Melodramatic, Allen Petterson opens up

The film is based on minimalism, almost Dada, + new aspects for education
/ science

Structure of the Narrative

Me - He

You- She your partner / the amalgamation, stilization of a Narrative
language is a matter of polemic.

In true deep therapy. Ends up in Aristotel's cave. Normal Oidipus.

Two gods, two religions, two directions of submittion.

After Shock feelings (You constantly search for an excuse, omission, or
an excuse that it's going to be better. That feelings of danger will be
left behind. People are willing to resort to incredible ends to reach
peace.

You can cherish bad feelings on the bases of your language inedequacy.
Things feelings should be named,

pinpointed so that they are understood.

I tried to borrow scissors - I was several times told that it's
impossible. I appologised.

Pertaning mindwork

1. that you eschew from eating something and you remember the past
experiences. -

During deep therapy and before after, - you should be warned that you
should eat such that yu are only

happy -at the same thime ypur behaviour has to be, or should correspond
with mundane.

1. I here think .- I'm interested in behavioral arythms / enclusively and
collectively called rythms -

possitive (inclusive exclusive negative).

I undergone a fast and in chill I recorded a song that was supposed to be
in geanra #rock more affinite to shanson, classical in Aretha Franklin

I hereby disclammer all possible notions and evocations that I use any
inapropriate versing (and hereby would call any such presumptions
inapropriate, absurd and prefabricated. I'm solving if surroundings and
instruments do tend to teach you almost manners.

I almost claim that harmonies -. and proper harmony stearing can lead to a betterment of verbal or

{speech functions. That's why we understand why education is so important.

I would be almost willing to theorise that a musician whu atteins a very good learning of ragas; {upanishadas, and mastery panatipata suta..

can better his/her speech (automaticaly, in succession), we is three, food. 1.Study of how to attein an

instrument

Theory.?/ People nowadays explicate on behavioral a-rythms with the possibility of reaching spiritual climax.

I claim these aspects impossible and call them illusional [disilusion, on e of the roots of evil / Buddhism.

125 pages the whole text..

---Page 140------------

On the Count of

Three

I tried several times to express profanities, that lead me to
understanding that I was in my fast cold and shaking. People with
belinguality have much greater perceptive and receptive knowledge

Behavioral extensive rythms [positive, normal, negative and based on
general good education, religious codexes of speech, behaviour and job
attandance. Person should be understood that he / she can live in a
mutual respect /according to religious codexes, good manners, education,
and the prospect of atteining a payed job, with + atteining at least one
instrument.

Problems arising from inundation

Ah, here you are! Said Penelope. Me and Harold were looking for you. We
stay on the other side of the ship. Room 108. The admiral told us we
should look for you here. Nice of you to come to meet us. Said Julian.
Would you care for a glass of gin in the lobby? Certainly, but that's not
the reason why we came to search for you. Penebrall wants us all in his
room in exactly twenty minutes. He says he arrived upon an ingenious
idea.

Julian knocked on the door of 104. After you. He said to Groine. And you,
little. He said to Cecil. You'd better stay here and guard. Let her come.
Said Groine. I bealive young Jackery is in there, as well.

Finnaly! Said Penebrall. We are waiting only for you. Harold beemed at
Cecil and Doctor Forlore brought her a chair.

I have to tell you what I came upon just this morning! Said Penebrall
with enthusiasm. He seems not to be alone. Whispered Julian to Groine.
Hush, hush! Said Penebrall. You'll marvel at what I found. Brag away.
Said Doctor Forlore. 7

10

On The Count of

125 pages the whole text..

On the Count of

Three

Three

--

I still don't understand one thing. Said Julian to Groine, as he plunged into his bed. What is it? Asked Groine. How come you ladies have a lot more fun then we! And he put his hands behind his head. Penelope is a wonderful woman, Julian! Said Groine. Did you know that she is a Duchess.

Julian was on the deck. Lost in thoughts as he gazed upon the waves. How far is Africa, Dady? Asked Cecil, standing by his side. Not far, Darling. About fourteen days travel. Are we going to see giraffes ? Who knows. Said Julian. Still deep in thought. Help! Help! Accident!

What's happening Dady? I don't know. Come!

Help! This guy is injured! Who are you, asked Julian. I'm just a passenger. Answered a fattish man in a suit. He fell down, all of a sudden! He seems to be in a terrible pain!

He is dead! Whispered Julian, as he rose from where the body was lying. The sun came up and something glittered under the dead man's jacket. Julian saw a dart coming from the man's back. He hesitated. Then crouched down and pulled out the dart. It was a small, refined tool. With red feathers. He put it in his pocket. 11

On The Count of Three

What are you doing?! Shouted on of the two guys in white vests, as they came. Nothing! Said Julian, agitated.

I saw, you! Said one of them. A tall man, not to play with. He seemed to be angry. The man is dead! Said Julian. The two health guards crouched down to examine the body. Come, said Julian to the fattish

125 pages the whole text..

On the Count of

Three

man. A scotch on me.

--

12

On The Count of Three

Julian was gong back to his room. He felt a bit tipsy. His mind just
started wielding back to the morning incident on the deck, when he heard
a noise of falling wear in one of the side rooms, used for storing
cleaning equipment. He began stealthing down the hallway, already silent
for a tomorrow ball. He came to the door of the cleaning room. Clash!
Cling! It seemed to him, as if someone was moving inside. He slowly
pulled open the door. A beam of light came filtering in. He looked at the
floor. A broom was lying across the room, along with an overtopped
bucket. There was something lying there, beside it. He crouched and
picked it up.

A wooden tube, just big enough to put in a suit pocket and stay hidden.
He pulled out the dart from the morning. Then he saw it! Glittering in
the light. Another dart of a refined making. This time with a blue
feathering. Is anyone here?

--

Just as I said, spoke Doctor Forlore, and everyone laughed. He was
jubilant at this night of such a merrymaking. Tell us, Doctor Forlore!
Said Harold, also in a good humour. How big is a difference

125 pages the whole text..

--Page 143------------

On the Count of

Three

between a Black Fever and a Yellow Fever! Well! Said Doctor Forlore,
depends on how serious both illnesses are. I remember when we were… 13

96

On The Count of

125 pages the whole text..

--Page 144------------

On the Count of

Three

Three

Afterword:

A versati l estoryofsuccess . Jill Bonmouth

AnIncredi b l e adventure,pl sagui d etonew Di n osaurprobl e mati c s.
Pen Guord

97

On The Count of Three

There was not too much I could do with the yogurt. I was staring into the
fire and imagining all the past memories from my travels. The cake was
baking in the stove. Another one, again. After a long spell of time. I
couldn't believe that Cecil came for the holiday to stay with us again
for a couple of months in our Shattau Mondo.

The maid came in and told me that the supper was being served by
Sharlotte. Oh, I didn't know that we had a special Occasion, I thought
for myself. Sharlotte prepared something I understood. Sousages with some
strange mushed porrage, or something that reminded me of foreign food
from far countries.

Is it spiced with some strange flavour? I asked Sharlotte. Oh, yes
Julian.

125 pages the whole text..

---Page 145------------

On the Count of

Three

It's a recepie I got from the work last year. Interesting, I said and bit
into the sousage. I was cutting the porrage with a fork and I understood
the logic of the Stereotipical Landscapes. I understood that Ajhan from
the Italian monastery was like the carrot and the sousage was like the
Tended Stone. Sharlotte grew uneasy about the supper. And She asked me if
I was alright. I wasn't sure and progressed through my meal. I was intent
on the carrot, that Ajhan would write me a letter that he received the
socks, or something. I understood that even him must play occassionaly on
the sand to look at sky. I would never realise that to cut the sousage
would be so difficult, when you ponder something as hard as a rock.

How far are you with the research, I asked Sharlotte. And I fidgeted with
the fork in the porrage to let it cool down little bit.

Sharlotte was in a good humour and she began to expound on certain
gnominy abot the logic of her work and her future progress in career. I
understood, I faked the sudden rush of needing to wash a couple of plates
and slipped the sousage under my sleeve. I went over to the sink and
tried to show how I could dry my hands above the soucepan when the
sousage slipped on the floor and shattered itself.

The current problems of behaviour as reflected by Stanisla Groff, the
founder of American medical centres, Erich From, To have or to Be, etc..

Tough isn't it. Said Sharlotte. What my work, I asked. And washed my
hands in the sink. Cecil rushed in and she was holding something I've
never seen before. A dragon box from dynasty Ipal Caqua, from Jipirada.
Oh, you scare me Cecil. Why don't you play with better things then my old
shoe box. Oh, what is it?! Asked Cecil and looked at the floor. The

125 pages the whole text..

---Page 146------------

On the Count of

Three

stone shattered in a way I comprehended. I slipped and broke the light
bulp with my hand and we remained in utter darkness. The telephone rang.
It was the Chief of the tribe from Kipaka (from the Bhutan

98

On The Count of Three

boundary where I got lost).

There you go again, said Cecil. Afterword

Tended stones on the themes of Stephan Culbreth are possible to coin new
theories in science of Palaeontology. Deep freeze theory is closely based
on the possibility of the extinction of Dinosaurs in the K/T boundary
through a heavy glaciation that took place and the oncoming precipitation
into soft tissue of extinct Dinosaurs. Stephan culbreth claims the theory
is possible through high salination.

99

On The Count of

125 pages the whole text..

---Page 147------------

On the Count of

Three

Three

Dr. Mary Sweitzer from Berkeley coined the theory of oscilotices taken
from the marrow of a fossilised bone. I claim that from my observation
taken out of very far countries in the World and ultimate travels into
France, Italy and Ireland renewably, there is a possibility of a so
called Vesuvius extinction in places that I call Steryotipical
Landscapes. Places of high vulcanic activity that migh lead into scenes
of high slaughter dipicted on the beaches and shores of Europe that
volcanic Pyroclast in the sape of volcanic ash acidic rains imbeded
Dinosaurs under heavy temperatures and they ultimately fossilised. There
is a possibility that through Iridium dating some tended stones might be
taken into Laboratories and taxonomicaly measured for a possible
existence of a whole DNA chain. I worked for the Robertson foundation
that still might send me money to purchase a Villa in France. I also was
in a close link with Albertov Palaeontological Facility and several other
professors even from the Kork, University, Ireland. Allan Petterson

100

On The Count of

125 pages the whole text..

--Page 148------------

On the Count of

Three

Three

101

On The Count of Three

102

On The Count of Three

103

On The Count of

125 pages the whole text..

---Page 149------------

On the Count of

Three

Three

104

Pi

Pi

Apple pie

Give me sight of distant sky

I was mumbling under my breath. The street widened, lights of the lampposts merged into a huge white-out. A ghetto prognoses..

125 pages the whole text..

---Page 150------------

On the Count of

Three

125 pages the whole text..

---Page 151------------

On the Count of

Three

125 pages the whole text..

--Page 152------------

On the Count of

Three

Belingual Organisation Crashed - Saurial stemps for collecting

We deem that stephan Culbreth coined an interesting phenomena in geological anomaly, and hereby we claim that the Europian sea migh have in prehistory been an interesting view of frozen shore landscape with large prehistoric birds swoosing down to catch its pray. The organs of prehistoric animals might have been carried into nests where they ultimately fossilised through high abundance of salt, calcite and chlorine.

2.

125 pages the whole text..

3.

---Page 153------------

On the Count of

Three

Benjamn Allan Schmidt

foreword

Here I would like to draw prognoses for orchestral music, multiinstrumental playing and ambient as a concoction for a storm. The fusion of classical instruments from India and Africa is my goal where I have to undertake a long journey for its reaching, There is generally recomanded very strong stress on training and exhilareting performance.

.

 Benjamin Schmidt

.

.

J ulian was hungry. As far as he remembered, he was always hungry. It had been six hours since he

crossed the border to Sierra Tralala. "Look at that dust everywhere, Molly." He said to a stuffed zebra toy dangling on a piece of string in front of his nose.

.The history of glacial maximums is a very palpable phenomenon which springs into current textbooks.

Stephan Mitten compiled a large prognoses for logical structuring of Glacial Maximums (LGM - last glacial max. 10 000 bp./before present). There is a wide history of researchers in Alps where thesis for cold life were prognosticaly put forward. The find of Alpine man with a batch of arrows can show possibility of existence of prehistoric man in Europe. The prognoses on which I base the logic for a palaeontological evidence is closely based on archaeological dating because it comprises the history of curren learning, The age of dinosaurs of aprox 65 000 000 years ago has shed light on the possibility of Europian sea and base evidence has so far been found of partial bones of Crataceous and Jurrassic animals.

.Half an hour and two minutes later he parked his truck carrying a cargo of metal rivets at a gas station by the

125 pages the whole text..

---Page 154------------

On the Count of

Three

main road. The sun was just rising.

.

.Julian jumped out of his cabin, his worn sport shoes tasting solid ground after an all night drive. That is, if

he did not count the short break close to midnight, when he was forced out of his driver`s seat to look at the moon. Julian wasn`t a character out of a novel. He was, after all, just a truck driver. His body seemed to move in good places as he scurried over the empty-for-miles road to reach an already comfortable shade of the corrugated-iron roof of the gas station. "Fries and a burger." He announced to the keen-eyed man, who stood unmoving in the doorway of the station as if he waited for this sentence all his life.

There exists a find which had been as an evidence for Early Jurassic life and its reconstruction has been successfully undertaken for the Museum purposes. There is a prognostic belief that the sea might have far surpased the logical thinking of current botanology and palaeontology with remaining traces of Precambrian life still visible on stones close to brooks and water gorges.

.Hungry truck driver Julian bit into his first meal of the day and wept.

._

.

.While a currently unimportant man, whom Julian at present was, had been shedding torrents of appreciative

tears over a theatre all alone in a land of hot chocolate, another man whose importance could not be doubted just entered a supremely important building on Mana Groul, Wer, and his mobile phone rang.

.

.The Wer-renown Museum of Natural History, located on the Rue de Mar street (because Wer had many

streets), comprised five interconnected buildings housing seven laboratories, and a mammoth library.
 permanent exhibition halls, research

.

. gSo why don`t we just say that it`s big?!" Shouted the very important man into the piece of plastic he was

firmly clutching in his right hand.

125 pages the whole text..

On the Count of

Three

.

. ❙gIt`s not that big, Doctor Forlore." Answered the voice on the other side.

.

. ❙gDo you mean to tell me that it`s not so important? I`ll tell you what it is … It`s the best news and that`s

the end of it!"

.

.Julian Stanton lifted his index finger on his right hand and, pressing a button on his mobile phone, he

interrupted the call. Julian was a handsome man. He had short, raven hair and not being exactly tall, he wasn`t small either. Neither was he chubby or boring. That`s what he was like.

.

.Julian Stanton looked around the immense entrance hall. It was early in the morning, but the sun was already

filtering through the high windows in sharp beams of orange light, dazzling the eyes of a scatter of early visitors, walking into the cool hallways of the museum.

.

. ❙gWer problems?" Said a voice, and Julian Stanton turned around. The voice belonged to Austen Parr, the

head of the department of Paleontology.

.

. ❙gParr!" Said Julian. "I have to tell you something!"

.The interesting phenomena of extinct man in peat bogs in Denmark is a prognoses for further research and

can trigger interesting phenomena in other branches of science, such as Palaeontology. The logics of the classical theses of fossilization can be for a while put aside as a prognoses for a type of fossilization called Calcification.

. ❙gLook, Julian. If you want to start persuading me about what I`ve just overheard?"

.❐gBut it`s the best news I've ever had!" Interrupted Julian.

125 pages the whole text..

---Page 156------------

On the Count of

Three

.

.Doctor Forlore took off his magnanimous spectacles and measuring Julian with his half-eyes till Julian

looked more like a freshly picked radish, he finally spoke. "Julian, if you need to take a couple of days off. You, see. Just to get out of here. I`ve always thought that you are a bit like me, rather seeing a Wer than Mer you have at home."

.

.Julian stared.

.

. lgAnyway, I`ve seen you! I think that you are a very wize man. " He finished his sentence and gave Julian a

friendly wink.

.

. lgDoctor Forlore." Interrupted Julian still, "But I was talking to Samon. The MD7 have found something.

They don`t know what it is, yet. It`s submerged, at least, ten meters deep under a thick layer of granite. The earth x-ray showed just a blur. I think it might be a Brontops, or even a partial Apatosaurus. They can`t judge the age of the layer, yet. Wer, Doctor Forlore! It's big!"

.Calcification is a process which can be explained by the classical Vesuvius theory of sudden heat and cold.

The magmatic rain which covered the whole Pompeys was soon after followed with a strong oncome of winter and snow-fall.

125 pages the whole text..

---Page 157------------

On the Count of

Three

4.

.Doctor Forlore made a long sweeping stare around the hallway and then looked back at Julian, as if

searching for an imaginary prop. He was getting old and comfortable. He was still charished, though, for his track with the latest scientific methods and goings, spending ever more time in his study with his gadgets or delivering lectures on the rudiments of Paleontology.

.‾gLook Julian," he said, searching for an excuse.

    .

    . gI have a lecture at ten. Speak to that new anthropologist from Spain.

    .

    .

.Doctor Forlore was an old man. He held a Ph.D. in Astrophysics, European Linguistics and a completely

new science dealing with predicting your own mood by the luminosity of morning suns.

    .

.Thus, while the completely unimportant, hungry, truck driver Julian was wiping his eyes into a smudged

125 pages the whole text..

--Page 158------------

On the Count of

Three

overalls and while a very important Julian Stanton still looked at Perr who measured him back in a slowly- filling museum entrance hall, Penebrall crawled out of his bed and looked out of the window.

.

. gWer!" He said to himself while staring at the gigantic ball of light. "I`m goning to make something good

for food!" He mumbled and averted his eyes.

.

.Doctor Forlore walked into his study. It was a room with walls made solely of bookshelves. He wasn`t

searching for a book, though. His hands were intent upon browsing through papers.

.Doctor Forlore was a Wer-renown capacity on many subjects.

.

. gWhere`s my Wer` medication!" said Doctor Forlore while frantically rummaging through the drawers of

his writing table.

.

.The pill had been prescribed by Dr. Gerard Monodin, one of a few people whom Docotr Forlore still trusted.

And although there was little reason in distrusting a person with such a name, Docootr Forlore nevertheless took his pillow of pure cotton diligently every morning to drink down with a cup of mint tea prepared by his maidservant.

.

. gWhat are you looking for? Asked Penebrall, entering his study. "My pillow! Answered irritated Doctor

Forlore.

.

.Doctor Forlore never married and Penelope had been the only woman who had ever crossed the threshold of

his capacious household. They had many things in common, one of them being their language, and the second their behaviour.

.

125 pages the whole text..

---Page 159------------

On the Count of

Three

. |gIt`s right on your table!" Cried Penebrall, seeing the big, white pillow glittering in the sunlight.

. |gWhere?!" Hollered a frustrated Doctor Forlore.

.

. |gOh Mon Ami, there it is!" Shouted Penebrall, crossing the room and picking up the pillow.

There is a possibility that when you exeed an undue heat over an object which meets sudden cold, the prognoses might in soft-tissue be put as Calcification

.Doctor Forlore put on his spectacles and observed the pillow - a big, white, glittering object in Penebrall`s

hands. Doctor Forlore did not believe in much prognoses, but taking the morning pillow was a rite he could not go without.

.

.

.

5.

.It was ten in the morning and Doctor Forlore was sitting at his book-overfilled desk. It might well be said

125 pages the whole text..

--Page 160------------

171

On the Count of

Three

that he looked serene, but inside of his skull thoughts of incoming problems were ever present. For the last ten years he had been trying to put down his memoirs, inclosing himself in his study for longer and longer and even longer periods of time to ruminate over his life and, of course, work.

.--

.

.The Wer-renown Museum of Natural History had a special building in the vast courtyard of its complex. It

was a ten story building known as "The Three Crown Hotel". The hotel did not receive

.its name for nothing. The Three Crown Hotel was notorious for its large and well equipped research

laboratories, lecture rooms, and also very clean microscopes. I mean to say, it all used to be. One day, the main part of the museum got simply overfilled with stuff. The corridors were lined with boxes of bones, back-bones, and bones that did not even look like bones. Until the situation got desperate.

. He dressed himself as a carrier, and humping a heavy box of bones, he rang the bell of the Hotel during a

lunch break.

.

. gI have an agent post to be stored in this building." He announced.

.Jackery was ready and he proffered a piece of paper carrying traces.

.

.

.

.

.

.

.Julian Stanton entered the Hotel and walked into the elevator. With all the boxes of bones around him, there

was just about enough room for him and perhaps, well, another box.

125 pages the whole text..

On the Count of

Three

.

.As he pressed the button to convey him to the ninth floor with the index finger of his left hand he had one

single thing on mind. He had to speak to Bertol, a leading expert on anthropology in the whole Wer and one of few people involved in the project in Sierra Tralala.

.

.The last two floors of the Hotel were the only survivors from the golden days, having still functioning

laboratories occupied by a scatter of scientists.

0.

1.The elevator door opened and Julian stepped out into a long corridor. stood in front of a coffee machine.

"Do you have any news off MD`s?" Asked Doctor Forlore in a tone of business.

2.

3.‖gYeah, I have them on the wire, right now. I just felt like a break, wanted my cappuccino, Stanton."

4.‖gIndeed" corrected Julian.

5.

6.„I appoledise, Mr. Stanton. Anyway, would you like to talk on the wire with our team?

The stigma of stalactite and stalagmite caves in europe is an interesting prognoses for the retainment of prehistoric small life. Such as beatles, ants and gnats. The logic by which we can follow the creation of a stalactite can be adaquetly prognosed as Calcification, due to a high Calcite satuated underground water.

7.‖gThat's why I've come'ere." Said Julian.

8.It was fifteen past ten and Doctor Forloret walked into a room full of people. Whether he knew if he

walked into a lecture room on the fourth floor of the Hotel Penebrall lived in a completely different Wer of his own. Still, he retained the dignity of a scientist and a frown of a scholar.

9.

0. gWelcome!" He anounced, and knit his brow. "We`ve assembled today in this luxurious room to talk

125 pages the whole text..

--Page 162------------

On the Count of

Three

about what life was like when the life that we know was not yet, but to
come. " He finished his sentence and looked around the room to make sure
he was understood. Then took a deep breath and carried on.

1.Penebrall looked around the room for the third time. "Today, my dear
all, we will talk about a creature very

much similar to us. We will talk about a sixty five million year old
dinosaur called…"

2.

3.Walrus and Baboon were sitting in the connection centre on the ninth
floor of the Hotel. A room full of

beeping, blinking and buzzing metal boxes in racks upon racks upon
shelves.

4.

5.Bertol was sipping his cappuccino while Julian Stanton, now expert, was
fiddling with the buttons of a

radio-signal receiver, trying to gain a contact with Sierra Tralala.

6.

7.BZzzzzzzzzz, Zzzzzzzzzzz, Whiiiiiiiii…

Spelologist can occassionaly be bewildered by the richness of natural
phenomena. The Ireland Natural Reserve which interlocks South Kerry and
adjacent Dingle Peninsula is an exceptional marvel of Natural Heretage.

8.MD7 was a cover name invented by Penebrall. Seven then stood for the
number of people involved in the

project.

9.Whooooooo, Whiiiiiiiiiii, Plummmmm…

0.

1.Doctor Forlore stroked the metal top of the radio-signal receiver with
the index finger of his left hand, and

the voice of Samon Mondaman filtered into the connection centre.
"Penebrall breaking news, Wzzzzzzzz." Bleated the loudspeakers.

2.A new day dawned on the Wer and a stray beam of sunlight tickled the
nose of Doctor Forlore. He woke up

as usual, with happiness. The door to his bedroom were open, and
Penebrall entered.

3.

125 pages the whole text..

---Page 163------------

On the Count of

Three

4.Thus, while the whole world was laughing and rejoicing, Doctor Forlore enclosed himself in his room and

knowing his time had come at last, he was illuminated with a one last, beautiful idea.

5.The news about a new paleontological find spread quickly. The evening issue of The Wer to Wer brought a

printed version of a Judy Tatters interview with Emerson Samon Mondaman , headlined 'Breakthrough!!', with two exclamation marks.

6.

7. llgWhat kind of book is it that the fossil`s holding?" Persisted Judy.

8.llgWe dunno. The letters are like we know now, if you understand. You see, like those we were taught at

school, but the language is foreign. We think it might be a dinosaur language.

9.II.

There is a possibility of bathing in Culdaf in the north West of Ireland close to Slieve Ligue, and the prognoses for thraveling by ferry to adjacent islands is a wodeful prospect of seeing an open ocean, lighthouse and the listen of tradinional Irish music with still spoken Gealige language before the night falls for contoling the influx and outfulx, unless accomodated in a comfortable inn, or a hotel.

0.Then it happened. Someone put on a CD with Mermer Jephers and his music filled up the dense

atmosphere. I don`t know whose idea it was, or how could I have possibly been forced into something so, but at that moment it seemed a good way of escapeing from Frederic Bruno. A human train of my co- workers was just passing by and a strong man named Bulvar Rolst pulled me in.

1.

2.

3.III.

4.My name is Julian. I was born as an amateur palaeontologist. I would like to relate to you a story...

5.

6.But first, let me tell you something about my childhood. I grew up..

7.

125 pages the whole text..

---Page 164------------

On the Count of

Three

8.Evelyn was my teacher. She taught me everything from my first letters and love for music.

9.

0.Rain was pouring down the Fitzgerald Avenue. It was four in the afternoon and all the lights in my room

were out. I was still standing by the window when the telephone rang. I went to the table and picked it up.

1.

2.Gerry, you know that I have not touched the guitar for the last two years!" "Tonight at Cafe La Pata, my

brother.

3.

4.We`ll send a het around. You`ll be good, you`ll be rich." "Alright."

5.

6.I .
 sky with rain.,

7.

8.He looked out thewindow. The rain was dwindling into a drizzle. Perhaps his lucky star was just coming up

somewhere in the misty evening, making the whole affair of my life more bearable. He switched on a little lamp next to his sofa with a jerk of the switch-cord. The clock on a wall was showing six. There it was. His guitar.

The prognoses for scoobydiving is suitable for Kilibags and the south of Kerry. The scoobydiving in Kilibags is done in an auspicious way as a suvenir in expensive boats and the prospect and possibility of seeing a whale with a professional instructor.

9.Elisa wasn`t on the scene yet, and Julian felt like he was part of a local music recording studio called

something like Local Metropolitan Orchestra. People in the neighborhood liked him

0.He played a wondrous solo.

1.

2.He was on his way again. The traffic was mild and he kept his pace steady. Thoughts were rushing into

125 pages the whole text..

---Page 165------------

On the Count of

Three

hismind. Things of the past. Evelyn once told him..

3.

4.Then he remembered, drenched as he was in that silly coat, an old mantra that Evelyn taught me on an idle

evening..

5.

6.⎮⎮gHow are you today mister Garry?" "Much better now that I have my sherry."

7.

8.⎮⎮gWhat`s new in the field of poetry if I may ask?" "A falling twilight, my friend. Dusk!"

9.

0.

1.The place slowly started filling up. Julian was on the look-out for who-knows-what. There were many

familiar faces and he would have sworn he saw Morton. People were drinking and talking loud. Many voices carried western accents and it was obvious that they came to hear the star of the evening. Mister Ian Gilliard. Yes, that was his name, and it was him that he was waiting for.

2.

3.Gilliard came on the scene. He was smallish but he seemed tall. He was dressed in white silk and his long

black hair was braided. This man never smiled and never ever had anyone seen him eating. He came to his table, deep grey eyes piercing through the souls.

The possibility of water diving in the south of Kerry is often possible for experienced personal teams with legitimate stemps and can be searched as a pass-time for young professional Scoobydivers. There is a special shop for diving in Dingle where in the summer you can visit a toorist information and watch small sail-boat races.

4.Before Mr. Gilliard`s glamorous appearance, he managed to get some information on his persona from

Garry.

5.

125 pages the whole text..

On the Count of

Three

6.Garry introduced the reading, as usual. He gave a short lecture on a book that had been around for decades

and finished up with a sincere..

7.Nightingale at night flees from shadows to the light…

8.

9. Awe stricken as he was, Julian missed to notice a sturdy gatekeeper who had been all the

time standing by. "What do you want here?!" He growled, measuring his skinny, bent figure. "This place is out of bounds for people who don't belong here." But Julian, averting his eyes from the splendor of the excavation works, was already far ahead of him. He now thought himself an archaeologist. "I work here." He said, and gave the bulk of a man a long and piercing look that made the man step back. "Go ahead then, sonny," stammered the gatekeeper. "No offence meant," and he made an opulent gesture as a token that his passage was free. Julian pulled himself up to his full height and reaching the gateman nearly to his shoulders he made a tentative step into an episode of his life that was to be both joyous and interesting.

0. In fact, it took fourteen days before all the necessary papers were signed and Julian was

officially accepted into the Palaeontological dig. The morning was heavy with anticipation when Julian was choosing a proper working gear from his wardrobe for his first day at work. He picked an old pair of trousers and a warm jumper the brown color of autumning leaves, for the summer was now long gone by and there were chilly days to come.

1.

2. Julian was instructed to take a pick and a shovel and when he loaded both instruments on a

wheel-barrow he set off across the Emmentalian surface to his prescribed destination. He was to work with two other boys approximately of his age, and when he arrived to the hole in the ground that was to become his home till the beginning of the winter, he gazed down and for a

125 pages the whole text..

---Page 167------------

184

On the Count of

Three

while savoured the ephemeral moment of his seeming triumph, for what he
saw in the pit was more than anybody else expected, except for rare
exceptions, ever noticed. He put on his spectacles.

3.

4. The dig proved hard, because all the fine work of sieving sand and
cleaning unearthed pottery was

reserved for women, and thus Isaac found himself working the pick from
six to eight hours a day. He wasn`t used to such labour and the first
three weeks he suffered greatly from a severe pain of all his
extremities. Yet he persevered and his sore limps had grown accustomed to
the constant shoveling of earth and picking up heavy stones.

5. Julian grew strong. Where there had been nothing but skin and tendon
were now muscles, and his hands

were callused and hard to the touch. There were many people working at he
excavation works and Isaac soon got to know them. Some were young and
pleasing to talk to, some were older with lots of experience,..

6. Julian worked from eight in the morning till four in the afternoon,
and when he came home

he read books of all sorts.

7. December came and with the Christmas at hand came snow. Yet Julian was
working as

hard as ever. There was a lunch-break at twelve o`clock every day and
Isaac would stand at a high table in the canteen, sheltering a cup of
coffee with both his hands and with an unceasing vigour exchange
informations about the newest finds. He felt himself an expert on
dinosaurs and although there were none to be found in the whole country,
after a coffee or a soda he would heartily start to expound on his far-
fetched theories on the eating habits and moving-patterns of these
prehistoric creatures that no-one remembered and few were erudite on.

8. The winter was gathering momentum and soon the ground would freeze up
and the

excavation would have to be closed till the temperatures allowed digging
again, which

125 pages the whole text..

--Page 168------------

On the Count of

Three

couldn`t be expected until spring. Julian was aware of this fact and with every day being palpably colder then the previous one he new the time was shortening. Yet he secretly feared the break - the winter would provide him with, because the dig had become for him a life- nourishing substance.

9.

0.
 ⚙

1.

2. The snow was falling heavily from the torn blankets overhead and Julian`s ears were prickling with cold

as he walked to work. He knew what was coming but there was something strong and unyielding in him that didn`t want to comply with the idea.

3. Julian entered the square of the Municipal House and the strong wind which had been persistently

blowing for the last three days suddenly changed into a gale. Isaac ran for the shelter of the archway but when he got closer he realized with sheer terror that the gate was shut. There was a sign on the door painted in red letters and although not saying much, still telling enough: "Excavation Closed temporarily!"

South of Kerry is a very sunny part of Ireland and the small mining vilage Trallee can serve as a pass-time resort with large spinning wheel and other children attractions.

4.

5.
 Introduction

6. Miocene thesis is an artificial name for this sci-fi book, in which I will be dealing with restoring of the

Miocene thesis with some new additions on the themes of John Demoor, and will analise phenomena which necessarily belong to my research. Stereotypical Landscapes, therefore is an artificial name for the purpose of building this book as an introductory guide for the beginning of drawing Palaeontological finds.

7.
 Allan Petters(on)

8. ..In

restoring the Miocene thesis, I built a small maketa where the logic of given

125 pages the whole text..

---Page 169------------

On the Count of

Three

prognostics was put in play. The resultant data led me to conclude the theories of the Stereotipical Miocene Microecosistems were possible to happen, or might have been omitted by classical thesis without any reprimands.

9.

You have a free hand, said the Chieftain. I'm going to the freezing device, it operates absolutely flawlesly...

0.There are ways and possibilities how to measure certain particles in rock, Penebrall. Said Julian. They were

standing in front of the Chieftain from the tribe Kipaka. Cecil was by Julian's side. Asking questions. Can you tell us who can allow us to dig in the vicinity of Patasalada? Asked Cecil. Patasalada is a forbidden place for newcomers and holidaymakers, said the Chieftain.

6.

125 pages the whole text..

---Page 170------------

On the Count of

Three

1.Cecil approached him and layed her glasses on the ground. The sun showed up like a shining jewl from

behind the morning sky and shown down with its beam at the box's occulaire. The spectacle chainged into a large buzzing clowd of light and a clawn spang out. Follow me, said the Chieftain of Kipaka. You can go

within. We have a seagerkin.

John Demoor's stones on the beaches – the possibility of a

fossilization of this stone is an

2.interesting way how to resolve the mystery of Tended stones. Said Penebrall. Cecil turned to the Chieftain

Ferber.Tended stone is a stone which underwent outside cause for its origination...
 Thank you, said

Julian
 You shouldn't stray into the forest. He said...

The Dingle peninsula can be considered as a Natural Heretage for the presence of a Dolphin in the local bay and a building specially built as a token of Natural Reserve Centre.

3. You

can bet, that we get, the machine set on - on the map.

4.

125 pages the whole text..

--Page 171------------

189

On the Count of

Three

5.

There might have been a wide possibility that birds in these dipictions of along the cold and snow environments might have carried the organs into their nests. I classify Tended stones further on Brass stones according to regular taxonomy and am of persuation that the glacial scenes in Miocene must have reached incredible magnitude. The organs might have gotten saturated with CaCl2. He finished..

Tended stones II.

6.The basic structure of Tended stones is sandstone, calcite and loas-loam.There are many places where

Tended stones can be seen and we usually exclude those that might have been carried there by hand in recent times. Harold was making notes.There is a possibility that some Tended stones might lead us to unreavel the possible prognoses for Stereotipical Landscapes which, nevertheless, are to be found in more remote places, like peat bogs, and reeds in close ocean or sea regions.

7.There was an incredible ovation in the lecture tent for Julian.

8.

125 pages the whole text..

On the Count of

Three

9.We also find claw marks on mud balls.

0.

1.He is incredible, Julian. Penebrall, I'm beginning to respect his oppinions.

2.

3.

7.

Said
 Harold...

4.I don't believe it Dr. Ferber. Said Julian. The possibility of this prestine absorbtion of calcite is beyond my

understanding. There is a more possible prognoses for a growth out of body in a shell for larger animals, than the possibility of such normal existence. I

5.don't think so Julian, I think your theories are horrid. I think you should more concentrate on a more

classical viewpoint and then only just start to come up with

6.some new possibilities and experiments.

Palaeontology in Ireland is taught at the Cork University in the South Kerry, and there is a bout two other

125 pages the whole text..

--Page 173------------

On the Count of

Three

colleges which offer similar education, as for example Magee University Dery, which nevertheless, is more specialised in retainment of the Language National Heretage and linguistics.

7.Julian was too tired with all this progress where Penebrall was headed. I wasn't sure

8.if the Miocene might have really been explained as such a powerful freeze on the

9.beaches. There would have to be marks on the stones and around the shores on the

0.local cliffs. You have to understand Julian that the Rocks in Miocene were really often of glacial origin or

at least glacially related. I'm sure you'll pardon me and will excuse Sharlotte to participate with me on this research. Said Julian. There is a strong probability that we might reach a better common understanding of certain facts with your wife. Said Dr. Forlore. The jeeps were coming across the sand, close to the cliffs.

1.Julian has always been a more of the seeker of the completely classical viewpoints on the Palaeontological

strata. He never was really sure where all this scientific progress was leading and who was supposed to prosper from these finds.

2.Sharlotte followed Julian into the Rols Roice and the car soon turned a curve from the Forest.

3.

4.that led inland. The excavations on the Beaches of Kipaka was a wonderful rest.

5.

6.The goal was to draft an approximate circumstance, and a hypothetical climate, under which soft-tissue

could have been preserved in such a pristine form. Julian went into his tent and layed himself on the bed made of hammered boards. He started reading a newspaper.

7.

8.

Calcification Fossilization

9.Dr. Forlore was talking to Cecil.

0. Jotto was a wonderful man of talents that was supposedly one of a few people that would ever understand

her. She was occassionally hindered from asking Greg if this was ever possible to happen. What Happen?

125 pages the whole text..

--Page 174------------

On the Count of

Three

1.Asked Hugh when he turned at her. This soft tissue on a small glass square, that was a sample from the last

traces. These are not so old, she said. There were other samples that got stolen half a year ago. Don't be so rash. Er, Maj was saying that the samples got stolen. The samples were sent to the

2.laboratory in Inkapa. The programe was unleashed for the possible findings of some logical conclusions in

Kipaka. I never new that you were so obstreperous in the

3.seclusion. I never was a miracle Maj. The

4.prototype of that Na Dana machine was sent from Jin to bar the space in the corner of the laboratory. I

would wonder if you could supervise it. Greg dissapeared to have a look at the jeep outside. He was a good deft hand on the chalcitron engine. There was supposedly a better way how to

5.progress through the repairs of these old chasts.

6.He scurried for something to put in Julian's hand, possibly a soda... . Hey? Who's there? Ah,

8.

Really? Don't say. Said Julian. I've

always wished you that, Dr. Forlore. Ah, but of course. Said Julian. I'll fly of tomorrow.

125 pages the whole text..

--Page 175------------

On the Count of

Three

7.I wanted to tell you that you'll be sent on bussines Julian.

8. The

Airbus turned on its Northward journey slightly into the left.

There is a possibility to speak to palaeontologists during Market times
and Fla(s), large aal country celebrations. A stall with a lecturer is a
good prognoses for retaining a small bulletin with erudite lectures and
can offer skeletal parts of long extinct Mammals and Whales.

9.It wasn't so easy as might be said. The woman, right after he came into
the door was working on something

to clean from the floor. Julian had a look into several rooms. He found
her at the counter with a bell.

0.She would have never noticed him if he wouldn't speak up. Julian
changed his mind and pressed. Oh, yes.

Mister. She said. Jane Ordbour.

1. I would like to sign my daughter for a course of Kapa research.3 Said
Julian. A couple lessons a month.

2.Oh, But of course. Just fill this form for our gnominy. Julian was
scrutinising the letter and begen counting

the prices. It was almost like when he and Jeferey Tatterstall were
skerminding the rock cliff in Keree. Such obtrusion he have not seen for
a decade. He ren under the rocky alcove and according to the blueprint,
they searched the rock massive where part of the shore went down towards
the water. Sharlotte, Jipi and Jeferey were putting the rocks aside for
possible scrutiny. They had a serious trouble persuading anyone to take
them downtown with the rocks. It would be100 money for a week with socks.
She said,

3.Jane Ordbour.

4.He came to the pub of Martna Prawl, a pub renown in these parts. He got
served so wonderfully that he got

a whole plate of buscuits for free. Julian was drinking a coffee. Marta
Jurba and her friend was sitting nearby and suddenly Julian realised that
through the talk, his wife Sharlotte might come to see him on a Jet and
he would have his daughter shipped here from overseas on a boat. Julian
was bursing for a while and then, tired of the market he crossed to the
closed hairdresser where they cut him

5.yesterday.

In Ireland, the current research of Palaeontology is devided into marine
and shore sciences, which is a mudane boundary for a country surrounded
with water.

Gigant Causeway, Ireland

125 pages the whole text..

--Page 176------------

On the Count of

Three

6.

7. Say what may, Said Julian. He scurried road with a hefty suitcase and entered the hotel like a prowling

cat. He slammed his businesscard on the counter of

8.

9.

0.

1.Amanda Jerennie and aquired himself to the hotel. I would like to stay in this

2.wonderful place. He said to Mrs. Jerennie. I would like to inquire if I can sign a voucher to stay in this

place for a couple of days with frenchtables like Mando, with my wife and my daughter. Is your wife comming? Asked Amanda. Oh yes. How do you mean to sign a voucher? She asked. Oh simply. I'll simply will sign a voucher and stay and then pay, later. I appologise said Amanda. That's simply not possible. Julian was eghast that his wife wouldn't even appear after such a long spell of loneliness. He stared at the Dipiction painting of Jap Gunar they had in the lobby room. Is this really Jap Gunar? Asked Julian. Oh, it is.. Said Amanda. He would market

3.it for a thousand money. You can accomodate Mr. Julian. Said the Landlord.

4.

5.Julian was lecturing..

6.

Here was a strong possibility that during glacian maximums in prehistory, just

like for example the last glacial maximum that took place approx. in K/T Boundary...

7.

8.
 Applause...

9.He is outstanding lecturer said Janette Gilinar to her husband.

0.

125 pages the whole text..

On the Count of

Three

1.Sublimining minerals might have seeped into the extinguished body of a Dinosaur and a mineral exchange

might have taken place. We find a similar comparison to some of the most renown Archaeological finds..

2.

3.It was raining, Julian didn't understand that he would

4.sleep in the rain when he was supposed

5.to get so rich. He ran to the pub, full of haste that Gorm Jopote would come to the hotel

6.in a Rols Roice to eat all the food.

7.Welcome home Julian, I'll give you a plate of buiscuites. But Borwal Rolst was amusing an old couple with

his talk from Jipakara. We ran down the palm tree forest. . I met him in a completely diferent place of the Wer called Barala. He was so full of stories that my collegue

8.and I started behaving.Tell me more about the secret, said Per, the woman. What secret, he asked?

9.How did you resolve the secret of that lost blade

0.from the palm tree. Oh, I have a lot of

1.repairs on the compost. Do you want to make it into this half circe with bricks?

2.Asked Mup. Yes I would love to.

3.

4.An old sailor, he is. Said the bartender. Full of stories. .

The zoo in Dublin is considered as one of the most beautiful zoos in the Western World and can house also a whale. It had been built by the help of Great Britain and is inernationaly awspiced.

5.Dr. Forlore was obnoxious to his reprisals. Julian did not understand that the Miocene theses

6.might really be plausible. Dr. Fereber was a

7.genious. He will one day receive a

125 pages the whole text..

On the Count of

Three

8.price for science, Julian. I don't understand how can you be so
impolite to his

9.oppinions. I just don't believe that Sharlotte is possibly to find
these glacial maximums of such a

prognoses, and I cannot be sure how you want to taxonomicaly measure
something that happened so long ago. He said The possiblity to

125 pages the whole text..

---Page 179------------

On the Count of

Three

9.
 p

lausible and new methods of investigation, said Dr. Forlore and the man
Gap pulled out the Gigant Sloth claw scimitar out of its sheath and
started polishing it with a special rug. The people entered theatrically
the Caffey. Mr. Ferber was sitting with his daughter to the right and was
talking to a large man called Jibi Mo. I have to tell you that the
progress of the evening fills my heart with beauty. Yes, said Jibi Mo.
The boy Roderic is a wonderful person with talents. My daughter is
studying Mathematics and

0.trace that the organs are even from Miocene might take decades to
measure and calculate. I dont't think so,

said Julian. Dr Ferber has come forward with very

1.Philosophy said Mr. Ferber. Jessica was just about ready and she handed
Jibi a hand for a mild shake with

smiles. The evening dimmured and the bend with Roderic began to play. A
boy called Pota Jurt was sipping his tonic and occassionally giving a
good

2.thump on the drums while Roderick wheezed his guitar to incredible
ends. Cecil was looking on,

apperently little bit dissinterested.

The Europian paleaontology can differ with the oncome of inland massives
and can candidate for different

125 pages the whole text..

--Page 180------------

202

On the Count of

Three

species of possible finds.

3.Footnotes of a palaeontologist:

4.A palaeontologist can be both professional and amateur. Here I'm more

5.concerned about the way I see palaeontology as an amateur performer of this craft.

6.As an amateur palaeontologist, you can obtain a licence for performing

7.palaeontology and except for fame, there's

8.nothing else can make you a

9.professional but perhaps a lot of work in the field behind you with still

0.little bit of that fame in your bowl2...luxurious places. Therefore I

1.would recommend a very nicely combed hair and

2.would probably encline myself to a black coat with very nicely polished

3.shoes. I found in the Borwla forests something that eludes the understanding of many. A tribe called

Kipaka7 and two more small separate groups who all seem to live in absolut harmony. I believe the stones in this region are beyond doubt part of the Lost World by Borwal Roist..

4.Feach ar un trad! Said Julian all of a sudden.

5. Cest

la vie, said Jipi. Qua La possibilite de comprandre la vie, Said Dr. Forlore.

But Julian wasn't listening. He set into his rols Roice and geared off like a lightning. Constant critiqu is a good way how to polish and sharpen the point of your own judgement. Dr. Forlore

6.Geophrey Manduno bit into a large cake.

7.Jup was showing his film. Jessica got all red pink and drank her

Dingle Natural reserve / Kerry

Fiction - Dna testing / feach look at the road

125 pages the whole text..

--Page 181------------

203

On the Count of

Three

8.tonic. She wore glasses already and Mr. Ferber

9.handed her a napkin. La tre Marvel. Said Julian. Everyone admired

0.the way Morton took care of his daughter that excelled at school with all marks.

1.Roderick was left alone on the stage. He got into a mood to fake tiredness in order to have a reach for a

lemonade. Cecil was able to free herself for this evening. Cecil was all smiles and everyone undertood that she was soon to understand the logic of the Kipaka tribe.

2.Sharlotte was standing at the bar with a large pink

3.fizzly water with an incredible straw and talked to a mysterious man.Comme sa, I never understood why

you don't tell Julian exactly what you want from your work. Connas ta tu. There different ways how to see Palaeontological cakes, and how to

4.approach it generally. You should understand that generally its almost a result...

5... I love it!, cried Girr Betor. I cannot predecess ...

6.What are you After, said Jibi Mo to Girr. Do you want to join for the

7.recital. Asked Jibi. He couldn't bear the possible weight of his work as a bussinessman anymore. Everyone

was clapping.

8.As Sharlotte had a wonderful talk on the Bulevard in the pleasant evening, Julian bought for a dime a

sherbet lemon to chase away thirst and had a look into the newspapers.

The south of France, such as for Example Larochell, is a good prognoses for seeing a whale and the biota is prognosed as slightly colder.

My concept of the theory explaining this phenomenon draws on a rather classical view of a quick burial. I think that the environment in the time of dinosaurs were, in fact, so different, and the layers that might shed some light on the floral situation are so deep in this region, that we are often left with just our imagination. Nevertheless, my inspiration draws on Denmark and the famous discovery of bodies buried in peat-bogs 4 I hold that food remains and parts of bodies might have been quickly submerged into soft environment. With no oxygen and the presence of preserving

125 pages the whole text..

On the Count of

Three

chemicals contained in peat, they might have quickly become good candidates on 'bog organs'. Then, with gradual change of conditions and temperature, the bogs dried up and soft-tissue fossilised.

So far, I have not enough evidence to either prove or disprove this theory, and I have no inclination to weigh on it if new evidence shows a different direction. Collective data C./S. 13.2. 2013

9.Calcification - A method of a deep freeze during

0.the K/T boundary.

1. Julian

Stanton..

There is a possibility of studiing Precumbrian life on the rocks surrounding beaches and small fossils of long-nonexistent plant and marine life can be found on the local beaches.

2.Wow! Said Julian to Sharlotte. I'm in the newspapers.

3.

4.A secret message came from Makara. Julian wasn't sure if this was possible to be happening.

5.He read three days ago that a woman from Delki suffered a sever breakdown and managed to earn a lot of

money in the Three Crown Hotel.

6.Her lawyer Murt Junava was all the

7.time by her side and writing notes. He had two calls and then they put Mrs. Shanan

8.O'deeny on Bursa. The market on the Marn was crowded with people and they began booming with glee

when she appeared to comb in a lot of monatery. George Whistling was standing by a tall man and had a call to the bank to make the transfer. Paleontological Meelie.

9.Mrs. Shanan O'deeny became the woman of the night and an immediate onslaught of

125 pages the whole text..

---Page 183------------

On the Count of

Three

0.camera press people that crowded the Hotel hall. Mrs. Shanan was a legend that no-one asked about several

days after. Julian read such articles. He always was interested how to earn on the market by investing several money.

1.

2.

3.

Julian was wondering where all the charitable people went, he wasn't sure if

this prospect of megalomany was the best

4.invention in the wer when all the prognoses should have been put to charity. This possibility was triggered

by

5.Dr. Forlore who was the only representative who was capable to quench the sorrow. She was the sole icon

of the project Kipaka, and Julian wasn't sure where to put his spoon. There was a lot of uproar about Frenchtables and Palaeontological bussiness Meelie that stemmed from the fact that someone migh soon get to succeeding,,.

The inland massive of Alps is a still very dubious place for studiing palaeontology, for the prevalence of snow, but Bergamo can represent with a better viewpoint.

6.Julian was walking here and there. He was very dissapointed in a manner that he hadn't been for a very

125 pages the whole text..

--Page 184--

On the Count of

Three

long time. This time, said Mr. Ferber, you have to understand that a new
Chieftain is about to spring up in the Makara legues... Julian wasn't
shure if Sharlotte was so obidient as to listen to Ferber when the
Miocene thesis was so close to be discovered and analysed. There was a
very powerful Meelie with several jeeps and a shevrolette as silver as a
hair pin and

7.Gaba Daka stood out of the expensive car with two guardians that
flanked her like

8.folage trees. There should be understood that you cannot dig or
excavate in the

9.vicinity of Parava to serve for the justice of the Conglominy. You have
been adviced

0.Julian that such obstreperous and longevity problems might result in a
postponement of your future career.

1.

2.

It's not my bussiness Gaba, said Julian. We love it here in French
Forests. Tres

beo 4 , We are serching for the possible findings of Miocene Landscapes
called

3.Stereotipical Miocene Micro-ecosystems. We may be analysing our finds
in the

4.laboratories later on, over the winter. Do you want to tell me Julian
That the Miocene micro environments

Beo /Life in Gealige, also good in French.

125 pages the whole text..

--Page 185------------

On the Count of

Three

were a necessary part of the

5.K/T boundary.

Sea quisine in the whole sea region Europe is almost logicaly intertwined and can be linguisticaly devided for sidedishes, and drinks.

6.Exactely Gaba, said Julian. There was a very powerful glaciation that might have

7.taken place towards the end of the pre-Miocene time and there is a possibility that the strong ice sheets

with the possible existence of very high uranium counts might have made even some new species spring to life and therefore also interbreeding and outside birth for diminished forms of life. Oh,

8.Julian. You are abhorrent, said Gaba.

9.You don't understand Gaba, said Mr. Ferber.

0.

1.Those are butterflies that Julian is so

125 pages the whole text..

--Page 186------------

On the Count of

Three

2.concerned in. He lied. We have a good basical knowledge already, to presupose big findings of incredible

lagre species on the bases of the Tended Stones Research. Do you want to tell me that the Tended stone isn't actually a heart or a liver but more of a Gigant butterfly larvea, said Gaba. Exactely, Said Julian.

There is a possibility to sleep in Scout camps (special mendatories for accomodating toorist for a cheaper price. Fumisino.) The prognoses for Paris is also relatively possitive, although the prices may be slightly higher.

3.We have a very powerful gnominysuppose that taxonomy and precise dating might always and every time

help us unreavl the mistery that every one is so scared about when he submitts his data.

4.You have two more months, Julian, to continue your research. Then we want you to diminish out of the

excavations and prawl back to your laboratories.

5.

6.Julian felt incredible happiness. People seemed to help him unreavl this mystery and he felt the Tended

Stone research and the ensuing Brass Stone research migh

7.percolate powerful results in the wintre labs.

8.I have to go buy a roll, said Jibi. You can, Said Sharlotte. Who's going to get you to

9.the town? Asked Julian. I will. Said Perr. I can get him to the town for a small

0.shopping in my Rolster. They were staring as Perr geared off with Jibi through the

1.dense forest.

2.Sharlotte was occassionally feeling very good. She was supposedly, according to the rest of the team more

concerned about her small notes rather then possible remedy to Julian's breakfast. There was a very good mood in the camp for the possible view of another several weeks in front of them.

3.Suddenly the wind slightly roze like foghorns from the dense high Pelorns and Jaka started screeming as

she turned around. Vu regarde! A large

4.Pelargonis of the raptorian kind

5.wherld the sky right above them like a large

125 pages the whole text..

---Page 187------------

On the Count of

Three

6.airplain. Run. Tranquil, said Julian. Run to the

7.cubicles...

Toorist manuals are a good way how to obtain bearings and a toorist office is usually very well searchabe through the internet.

8.How beautiful Cried Mr. Ferber and was stroking

9.his beard. Oh,. Cried Julian,

0.and ren to save him. The Meandropus Bird

1.leached down at Julian at a teriffic force

2.and Julian fell to the ground unconscious. I have to save him, cried out Sharlotte but

3.they held her back. I will erudite...

4.cried Mup Kondee..

5....huge horn from the tent.

6.

7.

8.Julian lay with his mouth half imbeded in the grass and he suddenly realised. Oh, Mondie, the

125 pages the whole text..

--Page 188------------

On the Count of

Three

9.calcification method might have been possible in the Peat surrounding
Landscapes that the Miocene wafted

with cold weather.

0.Sharlotte was helping Julian to his feet and they were all laughing.
The Stereotipical Landscapes. Said

Julian. They are possible at the Kipaka tribe in France. That's how it
is, Said Mr. Fereber. And wiped his forehead with a napkin.

Lecturing on palaeontological life is possible in museums and special
meetings. There is a wide history of modern lecturerooms.

1.

2.Julian arrived to the monastery through a bus stop right on a hill with
a small sprawling

3.town. He went up a little path and then begun circling a fence with a
couple of old

4.houses. He entered a narrow path that wound round the estate and got to
a tree that might have been a

hundred years old. This Ashram which name I won't disclose

5.was solely reserved for seven Buddhist monks that spent their time here
all year round. He arrived at

winter. An incredible time to travel even to Makara or Dabatasa.

6.Ajhan, Ajhan. Julian is here, Venerable Pranapata went calling. The
happiness and...

7.

8.Julian had an interesting moment in Jaral where he bought a toy for
Cecil,

125 pages the whole text..

---Page 189------------

On the Count of

Three

9.his daughter. Julian, said a monk whose name was renown in the whole
East and a part of Kipaka. He was

bettering the rocky bank that hemmed the upper part of the path with a
couple of nice purple flowers. He recognised him by his gate. Julian
hasn't been to the monastery for several years.

10.

Deep peace that began vibrating through him was so calming that tears
unstoppable

0.

1.went rolling down his face. Julian, said Ajhan and held his head for a
moment. He

125 pages the whole text..

---Page 190------------

On the Count of

Three

2.

3.came all the way up the garden to greet him. Ajhan lead Julian through the winding path and Julian forgot

tiredness with every step. This monastery was circled by large hills and fell into vellies and gorges with brooks of drinkable water. Come, Julian. You must be tired

4.with your journey.

5.Cecil might

11.

125 pages the whole text..

---Page 191------------

On the Count of

Three

6.have been left impatient in Rome for she wanted to come with me...

7.Ajhan

8.We set in the meditation hall and drank tea. Ajhan Karalata was sitting by my side.

9.He was my favourite monk, or better still I had a different favourite monk called

0.Maralata and Karalata was almost my friend. Me and Karalata simply enjoyed so

1.much fun together. He was the only one who understood me. I never was sure if

12.

.

2.Ajhan understood me as well. It was possible that he did, and was so far progressed in his search for inner

peace and holliness that he occassionally seemed oblivious to such basic logical moments. Karalata began to make faces. I understood that Ajhan was deep in meditation. Panatipata veramani..

3.Mak Jubi was interested in better prognoses for the day.

The interesting aspect of the geological anomaly is a prognoses for measuring Miocene in Stereotypical Landscapes. The regions can surmount water and prognoses for higher temperatures and humidity.

4.There was a lot of uproar as to who would first get to Switzerland and back without being ostracised and at

the same time being able to meet some

5.interesting monks from the Kapa Religion. The possibility about Gara's return for wintre labs was enough

erudition as to leave Julian still in a frenzy that he should soon finish the job with best results

125 pages the whole text..

On the Count of

Three

6.and come bak for home very soon. The resultant data were showing the Peatbogs possible for containing

enough silica and natural chloride were completely amiss. Julian did not believe his finding would be based on false data until Dr. Ferber came and said that the possible answer

7.migh lead into a completely diferent world. How do you want to explain the Miocene theses Julian without

being able to explain it through the peat. Julian was speeding as a lightning to catch up with Sharlotte in a state that something like

8.an obstacle was definitely anything but what he could currently afford. He parked his Caddilac close

9.to a pontiac in a small quarter in Switzerland..

The way we see sea-glass is a similar perspective by which we can logicaly surmount Tended Stones with further divisions. The Natural marvel by which water alters shapes of stones are visible also in Morrains in Alps, and water regions with a coder climate.

0.Julian generaly was never keen to rise eyes on public

1.places but the fact that he appeared in the hotel lobby with his hat still on was enough to disinterest

2.several mediators. I would like to make a call. Said Julian to Jina Had. The lobby interpretor and mediator

3.to put him through the line to his daughter in

4.Italy. He was sitting in the lobby room and thought

5.of Irna and Danam back in Ireland. Julian sipped his

6.coffee and tried once again to put this jig-saw

7.puzzle together. There was an interesting mat on the table and Julian was trying to read the advertisment on

8.chocolate. There was a time when he wanted to

9.invest into Frency Chocolate and possibly solve

0.the mystery of his future career, but he received

1.only a call and gratulations that he even wanted to participate. There was this sendstone crevice he

2.remebered that looked exactly like the kookee he got to his expresso and he remembered the

125 pages the whole text..

---Page 193------------

On the Count of

Three

13.

3.limestone from France. He wasn't sure if this would be possible. Cecil was put through to his receiver

4.and he got served theatrically with several biscuites with a poleve. It's not

5.in the peat, daddy, she would say. The peat is a false

6.and missleeding gnominy. Dr. Ferber came into the hotel. He was shaking, and sweating all

7.over. I resolved the mystery Julian, said Dr.

8.By deviding prognoses for the creation of a molded pebble and other structures with conical and musroom

shapes, we logicaly stimulate the ide map of regions.

125 pages the whole text..

---Page 194------------

On the Count of

Three

9.Ferber.

0.The answer is. And he put a piece of sendstone

1.paper on the table. Julian never knew his day. And

2.handed Mr. Ferber the receiver. Yes, it's

3.sendsone, Cecil, said Mr. Ferber to Julian's daughter.

4.You can

5.make the experiment. Through calcium and

6.chloride the Tended stones got saturated in nests

7.upon

8.sendstone bed rocks. So you don't think that there was a

9.large glaciation that took place? Asked Cecil.

0.There was, said Dr. Ferber. Because of Calcite and

1.Sandstone.

2.Julian never understood Dr. Ferber.The excavations were a wonderful possibility to acgnolidge that after

such a long time out, there

3.was a wonderful prospect of seeing something normal in one's hosehold. Gourn Jini was speeding

4.down the road to catch up with a wondeful evening and there was a possibility that she might soon

125 pages the whole text..

---Page 195------------

On the Count of

Three

5.attein enough knowledge to prepare tools for making Jelly cakes.

6.

7. There

is a possibility of escheving instruments in

8.a half-professional performance with a legible

9.result of percolation. The body receives the

0.stimulus from the past performance and the instruments vibrate in a
logical prognoses. I myself in my

There is a possibility that current onlook upon geology can bring an
idisputable balance into many aspects of its voccations and can help
furher to prognose different extrapoly in geological sciences.

1.study was interested in percolating a palaeontological Cake, said
Julian. Are you sure, Julian, that your

experiments are the best prognoses for a better prospect of seeing your
bussines thrive? I'm just an amateur Mer.

2.Julan woke up and did not understand the day.

3.The landscape was frozen and he found himself deep in the Miocene.
Large birds were swinging and

4.swooshing from the air and trying to catch its pray

125 pages the whole text..

--Page 196------------

On the Count of

Three

5.on the vast beaches. There

6.weren't any volcanoes as far as one person could

7.see and Julian found himself in the locality of

8.Western Europe. The large ocean was beating

9.

0.

1.against the frozen sea and the shores were

2.covered

3.with several metres of thick ice sheets. A large

4.prehistoric bird swooshed to the ground and

5.caught a

6.small dinosaur in its claws. The smaller birds

7.often hid the most precious of the pray in their

8.nests.

9.She sells sea shells.1

0.

--

1.People liked Julian in this region. And his

2.neighbours always helped him with so much. McAlice came the other day in her range-rover and brought

him a jumper. She didn't

3.know that he had already purchased four thick blankets downtown to brave the winter.

4.The whole thesis was a crux to a comfortable life. Though here in Dino, the summer resort for Karala

coming incessantly downtown

. An old proverb.

125 pages the whole text..

On the Count of

Three

5.for icecream, it seemed little bit too

6.biased. The rules are strict here Julian, you have to mind your job, Brun said to him

7.once. He was his closest neighbour, just across

8.the road. A mason by trade5, and an

9.Italian expatriot, he did wonderful teracotas with

0.Celtic motives and they became good

1.friends. I wonder what it is, this mutuality, Julian asked once half-himself, standing on a

2.2,45 mile stretch of beach with a boat perched in

3.the middle. We seem to be

4.understood only in locomotion.

5.Julian worked for a half-year in Kara. A city so

6.opulent that when he first visited it, his eyes got this strange wet feeling and his throat

7.remained speechless. He worked as an accountant for the local firm Hardis & Son.

8.Occasionally earning his extras by

The curiosity of a man with arrows can be superseded by the curiosity of a man with a sea glass and Tended Stones. The prognoses for this type of artifice is geological knowledge of nature and the knowledge of the intricasy of cave systems.

Auber - work in Gealige Auber whale[homework]

125 pages the whole text..

--Page 198------------

225

On the Count of

Three

Ther

e is a possibility that John Demoor was finding large stones that
resembled Dinosaurs and was keen to claim that such stones might be parts
of prehistoric animals. /I also am of persuation that Tended stones might
be organs of prehistoric animals that were subdued to sudden heat and
ubrupt cold. The stones are often found at old trees and tree stumps and
are usually distinguished in two types. One called Tended stone has an
oval shape similar to a small lung and contains veining similar to
clasical body organ veining. The other called brass-stone is similar in
shape to a small worm and its possibility to be a part of some body
system is strongly hypotetical. There is a possibility in spraying the
stones with polish for a nicer look, or simply retain the stones in a
box. A possibility of these two stones to be part of the Miocene
Landscape is probable, but not propped with evidence yet.

9.basking with a couple of musical instruments . Nothing mattered here.
All seemed to be strangely

obliterated by a bubble of timelessness.

125 pages the whole text..

---Page 199------------

226

On the Count of

Three

0. Julian

thought that some interesting aspects of Ire we completely put out of his

mind as he scummaged for a buiscuite..put his hat on and went to prowl down out of the excavations. There were stigmatising signs that some new species of plant life might be discovered and as he went by Amer Dubin's street he recognized a shop that was still open. He entered and begen choosing colours. He would decide to paint the Tended stone red and slightly yellow. He went out and met Amanda Deene. Bon Suar, Said Julian. What a splended night for a walk outside. Yes, said Miss. Deene. I was just about to go to the water and see to some things. Bon Suar, Said Julian again. He was very grateful that people...

1.

2.The ubiquitous prevalence of the Tended stones in mild and superfluous ground.. but you have to

understand Dr. Ferber, that the major find isn't the Tended stone itself, but more the Stereotipical

125 pages the whole text..

--Page 200------------

227

On the Count of

Three

Environments where these finds can be discovered. George Sir Parentoor once said that the possibility of finding something is strictly based on the wetness of your forehead.

Underground gorges and large cavarns can halp shed light on extinct and current bioflora and is closely linked with the basic cnowledge of the palaeontological sciences for the Europian gnominy.

3.Build your tent here. Said the owner of the country campsite just as a huge wave hit the beach downhill.

Julian was still in mid hight, because the beach ren down the slope and the inland started only just at his hight. He opened up a satchel and began eating bread and margarine. A whale jumped up out of the water and he heared several cries from the beach. This land was prestine, and almost without a touch of people.

4. Julian ate with the landlord and had a coffee. There will be a nice wind today, he said. Just a mild breeze

to occupy the senses. His mood roze with the coming of the early day. Julian went downtown for a breakfast and a cup of tea.

5.

You'd better hitch up the scaffold a wee bit, and then just put the lintel there.

We have to dig another five or eight metres to be able to analise the circumstances of a possible find.

6.

Maka was eating a sandwitch and occassionaly lifted a bucket of earth. There

was a strong possibility that we might yet discover something new in the Parafer.

7.

The Palaeontological excavation took two months and fourteen days. and we

found several Tended stones and Brass stones. There was a strong probability

125 pages the whole text..

--Page 201------------

On the Count of

Three

that the laboratories might lead us into some new discoveries. The possibility that Miocene was such a powerful change in the surrounding biota was an undisputable relevance. There were other problematics with fetching nutrients and a couple of people left for the Holiday.

8.

9.

125 pages the whole text..

---Page 202------------

On the Count of

Three

15.

14.

res

embles a coccoon. Said Sharlotte.

The possibility that it might really There is something I might tell you
be a remnant of a prehistoric Julian. I'm beginning to dislike the
butterfly still eludes my new theory of the out of body understanding. If
the stone will possibilities. It's just a butterfly, Dr. taxonomicaly
measured. You may Ferber. There isn't any logical find that it's out of a
completely inclination to deem that except for different strata. That's
exactly what the butterflies, you might ever find a

the

Julian found himself in an uplifted speed road above the city, and he was
gearing

like a lightning into a tunnel. He earned a lot of money on the stone
science research, and strict in thinking, asked

he was ever more keen around to realise the that surface. the Tended
Close stones might to tree serve Dr. as Ferber. a good Julian was
browsing result for the measurements. stumps If he proved and water the
Tended sources. stone really There belonged through is to the papers to
find out if he was

Miocene era. He might in future prognoses better understand as to he is
digging.

He tanked at Jan Da Rue, and the stones might have gotten byways.

They measured his speed as he went past. He is better today. Said Jaka
with to Perr, this. as He stood by the tree

he waved a flag with a tachometre surface, measurment unless device
they've in his hand been and scrutinised made a the stone that took

There were several the beginning. the City and

nobody was sure who was to solve the mystery of the Tended stones first.
Kapa Dava had an outstanding speed machine of Perevoir steel, with a huge
motor spitting flames. The Para Gana had an inceredible red car that
looked like a rocket. Everybody new about the research, and nobody was
shure if the stone might really propel a car, but mor likely the science.
Para Gana roze in the morning and made herself a coffee. She liked herbal
tea more, but she remembered the theatrical performace of Kapa from
yesterday, and did not feel completely well. She

her car. Anything new? Asked Jope? Her landlord was helping her with a helmet. Nothing special Said Para. Her car was called Mackee. She sped off and dissapeared under a small bridge.

I'm suggesting, said Julian. That living dinosaur around. Is the

stones are constantly to be found so

still no logical explanation where still reasonable enough to come up

how progressed like speedster through several all

check on his note pad. there put right in operating people in his interest for such a long time.

125 pages the whole text.. dressed herself into her special neopren clothes and got into

---Page 203------------

On the Count of

Three

1. There

is an interesting western possibility for fusing African, Indiand theatre

into new prognoses with stage design and the building of theatre. I personally incline myself to interesting prodelemas in theatre props and sets with orchestral music.

2.

Then it started raining. Such a Farine and splitter Julian hasn't experienced for

fourteen days and with the roar of a thunder the motorway darkened. Julian stepped into his Rols Roice four weel with a holstered front and sped like a lightning down the road. There is a strong prognoses for a better prospect then finding a stone which

3.

Julian got into his Rols Roice and turned on the ignition. He felt like buying a

sandwitch and then going for the excavations again.

4.He pulled off at the excavations and put out his umbrella to serve him as a cane. He was feeling better

today.Can you hand me the chistle. Thank you Mer. And the spoon. I was wondering what the evening will bring for food. is going to be nothing, but lentils. Ah, lentils Said Dr. Ferber. He was more interested in the excavations then food. Julian, there should be more of the earth should go to that heap. Yes, said Julian. , me what it is, Julian. Said Dr. Ferber. This absolut devoutedness to your voccation. You get such respect, that I occassionaly am weedled out of logic. Mare si, said Julian. He was more interested in Dr. Ferber out of body presumptions, then before. Dr. Ferber was a keen hand on budget, and his logic for Palaeontology amouted to occassional questions. He was often stricken by an ingenious idea and never was he able to finish up with the mundane.

125 pages the whole text..

--Page 204------------

On the Count of

Three

The freezing box is ready, once we find a

whole species, we can sustain it.

5. We just correlate the given data, said Julian. You cannot possibly presume that any new idea will be

immediately put into action. I think that your suggestions are interesting enough for concluding that the Tended stone can really be a butterfly of a larger species. Said Julian. Was the Miocene so interesting a relm of cratures? Asked Mr. Ferber. It was, said Julian. After the dinosaurs, smaller mammals exchanged the larger animals and a completely new life began on Wer.

6. So, you think there isn't a possibility that the butterfly might yet fly out into the sky? There are ways how

we can reconstruct the past life from the samples and finds we discover on the excavations. We analise the

data and draw logical prognoses given on our judgement. How

many stones have you found that

you think are exceptional? Asked Mr. Ferber. Exceptional is each which grows onto you, said Julian. There are stones of quarts, or schist, but also harder stones of different origin. Most softer rocks have been made by impacking of sediment, while some extrusive rocks like basalt are ususaly from deep within the Wer. Do you see any extrusive rocks around? Asked Dr. Ferber. No, there is just this loas-

125 pages the whole text..

---Page 205------------

On the Count of

Three

loam bed we find ourselves in. We never carry on much further, unless we find something new.

7.

8.

Thank you. Said Dr. Ferber.

Delivery, said the letter. Oh, Mon ami Julian. I understand that you are bringing me the cake! You can put it in the freez-box. Where are the rocks? I'm gonna put them with the socks.

9.Julian, here is your ambrella, Sharlotte told me that she will wait for you in the Shattau. Said Jipi.

0. We have problems with food supplies.,said Jin to Julian. Julian understood. He went over to Maka, and

told her to call Demin from downtown. Tell him to bring several crates of food. We will freeze it in the sendstone cellar with artificial ice. I understand. Said Jin. Hall it up! Cried Mak. And they began pulling the pulleys. Look, Cried Jipi. Its such a huge monster. Julian roze with dignity. He felt full of happiness after such a long time of anticipation. They pulled the opalescing creature into the air, and Mada Jurba played a pozone and Jipi a trumpet. Julian shouted a word of command and they heaved up the huge skeleton up and up. It's enough, said julian. Not bad for the present, said Dr. Ferber. We understand that you are winning this year's competition! Sharlotte pulled up with a pink caddilac and stood ajar with surprise at the sight of the Gigant opalised Skeleton. What species, do you think it is? Asked Dr. Ferber. Looks like Gigant Diplodocus, said Julian. How are we going to get it to the museum? Asked Jipi. We will hall it there. Said Julian.

125 pages the whole text..

--Page 206------------

On the Count of

Three

1.

The orchestra played, and through the pomp and prestige, Julian stood on a

chariot with a raised platform as they halled the Gigant Skeleton through the town. Everyone was clapping and happiness was seen in everybody's face. So be it, for ever after. The skeleton found its use in the museum. And Julian had a day.

2.

Text book for modelling

Dinosaurs

There are different ways how to model Dinosaurs, and how to approach modelling generally. You have to, or should, understand that modelling generally is almost a result of drawing, [figure drawing, figure composition of mumanoids, still-life, even painting, like oil painting, acrilic colours painting, and aquarel].

There are different ways how to approach figure drawing, according to your abbilities and possible inner, intrinsic inclinations.

To sign on a drawing course with the ability to approach a woman/man model is one of the best thing you can do, unless you want to be destined to painting and drawing according to photographs.

1a/Footnotes- You should understand that drawing according to photography, in the real onlook upon modelling, should serve you only just when you've mastered drawing completely or for a very deligent, and painstaking study of human body.

125 pages the whole text..

--Page 207------------

235

On the Count of

Three

It may come as surprise for you to do so, but out of a photograph, you can never reach such results of drawing a logical body, structure and mainly 3d roundiness, so important to be understood, and absolutely mastered in, or for, 3d modelling (plaisticine, or even on 3d programes).

2b/ Footnotes

2. It is important to study figure drawing for many (many, many) years, for you to even start to comprehend your own logical goal in your slowly mastering craft.

The constant supervision of a professor, or a lector, is one of the most fundamental, and can't do without, attributes for your successful passing the course and understanding in your own logic your own limitations and possiblities.

There are, after mastering pencils and coloured pens, possibly also aquarels, or even at least the main branches of printing methods [lithography, dry needle, etchings and copperplates, linocut, woodcut.

1b/Footnotes- You have to, or should understand, that there may not be enough space for you in your study to fulfill all the printing tekniques to such wideness, At the same time, it is strongly recommended that at least a brief overview and some rudimentary knowledge of the bases of all the printing teqniques should be mastered.

Also, occassionaly, it's better to ask if you can try out the appropriate printing method rather then sublimin yourself upon unwantedness, without any logic.

There are programs that have a very good use in 3d modelling and will cite them here.

3D Studio Max, Maya are one of the most expensive craft 3d animation and modelling programes on the market, and its not only possible that you may never even rub off on them, or meet them in your career, but

125 pages the whole text..

---Page 208------------

236

On the Count of

Three

also that you may not evere even feel the need to be part of them.
Reccomended programes are Blender in such case as this is an ORG
programme (designed as freewere) but with the possibility of needing to
obtain a licence on it anyway.

I here talk about 2 programes available for you that you can easily
borrow from your University and after mastering even purchase under the
supervision of your parents or very good friends / or university
professors who may further instruct you or hold you in the program
keeping and maintanance.

One of the programes is a Gif Animator , a wonderful tool for taking
snapshots of your 3D model (possibly a dinosaur) and putting it in your
computer.

2d/Footnotes - you should understand that currently there is so many
freewere programes available for animation that under the advice of your
professor, or perhaps with the investment of 10$, you can obtain an
enough good programe to justify your work for 10 or 20 years.

When Animating in Gif Animator, and if you really want to enjoy the
truthfulness and the imprint which plaisticine animation leaves. You
should photograph the dinosaur in real animation. In otherwords, move ad
animate the dinosaur on your table and then load up all the photographs
into your computer, into Gif

125 pages the whole text..

---Page 209------------

On the Count of

Three

Animator and just run the animation. Gif Animator offers the unique possibility to animate inside of the programe, I would nevertheless say, that by making just one photograph, and trying to animate in Gif Animator by means of the tools is not recommended, unless you are an outstanding drawer, drawing practitioner.

2c/Footnotes- You should never get bored or tired by the clumsiness of your work or your program(s), because such feeling stem from your own lethargy and the feeling to devote time to something else. Then, you should probably understand that you are not a scientists, but a person who wants to study something else.

Education is one of the most important things we can get, and obtain in our lifes, You should therefore choose deligently branch of knowledge you want to devote your time to.

125 pages the whole text..

--Page 210------------

238

On the Count of

Three

The other programe I wont be even mentioning. The relative profusion of free software, should suffice you in your several year study and practice.

Drawing shapes [drawing still lifes, like a Jar mugs, apples, onions and drapery, possibly built by your professor] belongs to the most basic and rudiment maintanance for your real life 3d modelling out of clay or plasticine.

The Advantage of Plasticine over Clay

The advantage of plasticine is its unnecessarty altering its state, as in baking in a kiln in pottery clay. You just buy a kilo of one or multi coloured plasticine and start modelling on a wooden stick, ball or a cashired paper skeleton. The plaisticin / or a straight model out of it does not need maitanance and imagine that a clay model of a rhinoceross folls to the ground, the plaisticine model has a much greater longevity.

Possibilities of Modelling Dinosaurs

There are several possibilities how to model dionosaurs. As I said at the beginning. Drawing (meticulous study of still-lifes and nudes (nude drawing is quite common in art lessons in the Cyech rep. and other western countries. Where the history of this paraphernalia has a relatively deep rooted tradition Frantisek Drtikol photographs, Saudek, Tono Stano etc..), and then averting your work towards the animal life. The study of finished illustrations of animals out of texbooks etc.. Helps you reach a firm hand in averting, and

125 pages the whole text..

--Page 211------------

On the Count of

Three

transmutating drawings into 3d real models. (real life is a different word meaning almost as tall, or big as in reality).

The potential of your education is a hefty friend in teaching you enough patience to stay at a modelling board for dosens of hours. By the time you become a modellar, you should be almost aware that you are giving up on illustrations and might have still the ability to illustrate a book. Some modelling craftsman never reach enough talent or stamina to actually ever illustrate, but with this aftermach they can become outstanding craftsmen of relatively large models of animals close to a sculpture art. But still retaining its scientifically purposes.

You can, in modelling (dinosaurs or animal life), follow several steps that differ right from the beginning. From the conscious use either cououred plesticine to choosing amonochromatic brown. You can start with modelling via, or throug the use of your hands and fingers to the resort of small spoons and similar woody instruments taken from pottery. As you progress through work, alog with certainty you can adopt new instruments or simply work with broken ones as you'll find them much coasier in use.

Plaisticine is a wonderful tool for modelling highly recommended for scientifically methods and in possible later taking of casts (a cast).

When preparing a cast simply use wood or even plaisticine itself, and never expose eny of both to fire.

(plaisticine is flammable, Beware!)

There is a nice time

when chalky material dries up, and you tear off the wooden, or plaisticine cover (two or three [occasionally multiple] forms. Molds]. You should solve that the risultant cast or outprint of the plaisticin modell often has to be further furbished, worked, painted, polished until you gain any satisfactory result.

In this textoob, we are engaged only with the sidig of plasticine as a final result, which realy occassionaly can [under good handlig] be exhibited at places, simposiums, or simply school windows.

125 pages the whole text..

--Page 212------------

On the Count of

Three

In the history of the Palaeontology, Antropology and several other adjacent sciences, there have been people abroad, who won fame through sculpting large tex-book creatures according this way of model-creating.

Res is an important par of work. Exercise yoga under the supervision of a learned trainer in your free time.

Tyrenosaurus Rex, Diplodocus, Rhynoceros, etc.. There are many ways how to approach modelling such animals, and the animals from our prehistory well.

Scrutinous study and reading of coloured, illustrated texbooks about about shore, and marine life, should fill a lot of your free time, unless your voccation can never reach enough gnominy, and you'll soon lose a good hand on your work.

Constant critique taken out off illustrated books, can in your later age and career be become a good friend of yours, particularly f you are interested to work your hobby in a freelance way.

Computer programs, aren't so important if you are good and talented enough to earn a colegue by your possitive reputation. Your work has always need to have a submission source with a possitive feedback, otherwise you might be adviced to put your voccation down and care for a different job, and give up on your talents altogather.

The difference between a good artist and a mediocre one is in the luck and charm of obtaining commission work / given for pay. Therefore you should still see this adjacent extrapoli to science as a good hobby of filling up your time, unless interested in sport or teqnique etc..

125 pages the whole text..

--Page 213------------

On the Count of

Three

Possibility
 of
 retainment
 and

renewable of netural reserves.

I spent a log time in Irelan (see. Listen to my lecture 004)/ Erasm -
Maggie University Dery 2005, 2014/15 then particularly Co. Kerry. Cork.
University.

There is a powerful belief that most natural resources and reservations
are in plenty and lush to the overflow. Nevertheless, as a scientist
interested in Nature, be in maybe more amatier, rather then professional,
I would theorise on the possibility of new Natural Reserves arising.

You have to understand that according to current trends of fashion and
landscaping gardens (private and state owner places), we reach the point
of common cognominy where we can claim that the fashion for now and
possible oncoming years will lead us into Ecological homes [such as we
may know from England] and ecological landscapes, new [possibly called
artificial forests], forested parks and landscapes.

There is a tangible inclination in people to like linking and connecting
to old and rut-in knowledge and places, just as the History often teaches
us.

There may, in the prosperous future, arise a fashion for this type of
liking. Old sacral hamlocks and the surrounding area + the seeing these
habitable places renuated according to the modern thesis, is a wonderful
opening and a new hope for revitalising dishevelled places into
reservoirs of lush plenty, and forestation of possibly endangered corners
of forest.

There may be a tendency in my just partly intaken oppinions, that I
nevertheless don't put so much interest or a firm value to. That, we in
the west, often become a seduced observers of changing vegetation strips.
These fields of vegetation and large stretches of green-foliage trees
such as Oks, Lindls, Birches, poplars, but also willows that hem the
shores and banks of rivers and slightly wet ground strips + also firs and
some coniferous trees such as, might one day turn into guarded vegetation
strips where lusch nature and its value

125 pages the whole text..

--Page 214------------

242

On the Count of

Three

and welth will be further enhanced and protected.

There may be notions that stem from legends, as to how the Giant Causeway was created in the North of Ulster in Irela. To see such a place protected from the often impact of onlookers and visitors is a wonderful prospect for many western countries to resolve and understand the current fashion change toward netural preservation current.

To visit Prague Zoo in Autumn just as in a full bloom of the spring,. Is a lifelong moment for a lot of visitors and happy seekers of natural beauty. There are several places in the Prague Zoo which by its beauty won renown all over Europe and even on some very important international meetings. Dolphins, Seals, or even ducks bask in the late afternoon warm sun in spring, and as you change direction according a given map or just following leading arrows and signs, will soon see tigers Bears and lions.

Shipanzees and gorillas belong to on of the most beautiful atractions that hardly any child will forget.

The Zoo in Dublin is a very similar one just as the zoo in france. It is possible to say that as mentioned previously. There might be a certen profile that some vegetation might be strongly influenced by current trends in changing rythms and a powerful influx of incomers into western Europe.

There is a very strong belief inmy thesis that the south of Franc, just as Ireland have seen a certain alteration towards the sub tropical climate. Palm trees often brought from vaccation, or bought on the maret thanks to current fashions and the possibilities of international trade, we become vitness of ever more hoseholds adopting the palm tree in the garden trend and behaviour of caring. There might be a possibility in the future under the current logic of the trade and citizens vacation trend that the spread of exotic flowers and bark- woods will further continue.

The proliferation of garden seeds into the open landscape has not yet been marked by me as obvious or, intrinsic upon any impact on the current vegetation. At least insomuch that I would claim such notions overstretched.

The often almost superstitious tug in meny feshion-influenced people towards the prestine forests and Rain Forests of Africa, is an interesting factor when juxtaposed with the might be future trend of returning

125 pages the whole text..

--Page 215------------

243

On the Count of

Three

towards old velues ad its logical, straight heritage in the West.

People from these countries (such as France, Italy, Cyech rep. and Germany) often invest a great deal of many to undergo excursion trips, either funded by their adjacent school, or daring on their own, into far-off countries of the Ashia, and Africa, South Brazil and similar sub tropical, to almost tropical leanage. The curiosity of these young travelers might very soon lead in fashin, and according to me, maybe particularly seen in their procrastiny, into a certain comeback towards our trial [possibly Celic] heritage.

I would see this fashion as being well reflected among the travelling people who gladly afford themselves to understand the incredible tremmor and obligation in adopting foreign ways of thinking. Therefore, I think that in a couple of decades, we might possibly even become a witness of the young generetion teaching their procrastiny about the value of their (or our) value of our historical heritage.

Afterword:

A versatile story of success . Jill Bonmouth

An Incredible adventure, plus a guide to modelling Dinosaurs. Pen Guord

3.

4.

5.
 Monastery

6.I`m communicating with a proffesor from the local University. He says that he

7.is beginning to understand me, but he is not sure about the prices.

8.He says that I should study more if I want to be justified for my work.

125 pages the whole text..

--Page 216------------

On the Count of

Three

9.I was asking him if he would be able to give me a doctorate for the treatise. I

0.know well enough

1.that this is close to inpossible, but then I know that it is something not completely

2.out of

3.question, is he reclines to my ideas. I could get an honorary doctorate, Ph. d. c. (theoretically), you know

what

4.I mean, but

5.then I don`t know if one would be able to carry such an honor, at all! That`s a big

6.thing to have,

7.you know! The last several days, I`m basically pondering if I`d like to have

8.anything, man! It is a second time I roused that man from sleep.

9.I was also asking about a price for science (for you).. They are simply completely

0.oblivious

1.towards any action.. I`m feeling like a big, big break!

2.break news as soon as I can:
 Market

3.

4.I had to see it in Paris. There was just a little time after breakfast in Lune de Mon and Jenifer Alber was

already on the line with Mark Brach to fetch him good prognoses on the Bursa. was staring at the newspapers and imagining my past travels. Did you leave the keys in the door? I asked. I locked it twice, and then one more. Market, said Jeen Oerr is the most stipulating thing in the current time which can assist in many difficult occassions also in charity. There is a powerful gnominy that the skeletal head is opalesced beyond table prognoses and ruby and schist opals change into milky achate. The beginning price is at 1500 money and will grow.. Two.. She was calling next to Mada Anne and the bussinessman who seemed to be a more of a caller was on a line with the marketeer. There was a powerful desire for the object.. They put Pet

125 pages the whole text..

--Page 217------------

On the Count of

Three

Gerr on Bursa. The auction room was heavy with raised hands. Five... 2500 money..

5.

6.I will never understand the prognoses said Leena to Julian. Me neither, said Julian. I feel almost immune to

that Perry man.... Said Sharlotte That was interesting of Doctor Forlore how he had it? Said Julian.

7.

8.

9.
 Science

0. Dr. Mary Black has the term oscilocites. Carol Janings.The Tended stones II. On the themes of John

Demoor is possible to coin new theories in the science of Palaeontology and brings an interesting view on the Deep Freeze theory. There is a possibility that towards the end of the Miocene K/T boundary a strong glacial maximum might have caused, or better still had a smaller impact on surviving species of dinosaurs. I am interested in reconstructing the Miocene thesis in a healthy manner which might bring elucidation on a lot of yet unanswered questions. John Demoor is a strong adherent towards the salinity theory, but I still tend to understand better some logical aspects of this problematic through snow and ice and high temperatures.There is a possibility that through Iridium dating some tended stones might be taken into Laboratories and taxonomicaly measured for a possible existence of a whole DNA chain. I worked for the Robertson foundation that still might send me money to purchase a Villa in France. I also was in a close link with Albertov Palaeontological Facility and several other professors even from the Kork, University, Ireland.

1.
 Allan Petters

125 pages the whole text..

On the Count of

Three

16.

2.Julian PS: The last find of mine has unleashed an incredible meelie in the Palaeontological circles.

Soft-Stone Landscape

3.

(a stereotypical micro-ecosystem)

4.524.

5. Me and Steve Culbreth are trying to expound on the fact that not only it is possible for remains of soft-

tissue to get preserved on fossil-bones1 (dating back to the times of dinosaurs), but we would like to show this phenomenon in a completely new light; bringing down collective evidence based on years of observation,2 and field research, that whole organs of prehistoric life have been preserved up to the present, and hereby illuminate something that we collectively began to call a 'Soft-Stone Landscape' (A Stereotypical Micro-Ecosystem).

1 "Soft tissues are preserved within hind-limb elements of Tyrannosaurus rex ", (source: Schweitzer, 3/25/2005). 2 Steve Culbreth has been examining this phenomenon on his own since about 1990.

125 pages the whole text..

--Page 219------------

248

On the Count of

Three

6.

7. The following treatise enlarges on the palaeontological discoveries made by Stephan Culbreth in El

Granada, California, and me, Benjamin Schmidt in the Czech Republic.

8.

9. In 2005, A scientist, Mary Schweitzer`s team, published in the journal Science news that raised commotion in scientific

circles. Her team, after dissolving a newly found Tyrannosaurus Rex`s bone in a solution that completely disintegrated the

fossil matter, discovered actual blood vessels, bone matrix and osteocytes 3 .

0. Stephan Culbreth lives in El Granada, California, and for more than a decade studies finds that retain traces of

dinosaur habitat; stones which he himself began to call "Mystery Stones", and I gave them the name Brass Stones A , and Tended stones. B

1. The rocks carry resemblance to fossil organs. No matter how this may sound if compared to the Mary Schweitzer`s

discovery, me and Mr. Culbreth now claim this fact to have a scientific significance.

2.

3. Theory: Stephan Culbreth

4.

5. Stephan Culbreth builds his theory on the idea of high-salinity in water and neutral buoyancy. He thinks

that stones that carry marks of treatment by dinosaurs, are actual remnants of food; residuum, or simply; left- overs from prehistoric feasts. They carry semblance in shape and their rock make-up.

6.

7.

8.He says: "Food (meat) was cut squarely, into small pieces. 'Predators', like raptors, had smaller stomachs.

They butchered their pray rather than swallowing large pieces at a time. Contrary to herbivores, they had much smaller stomachs; absent of gastrolithes. What we find are fossilized remnants of food."

9.

0.

1.

2.

Stephan Culbreth

3 Robert Lamb, How can soft tissue exist in dinosaur fossils?, http://science.howstuffworks.com A Brass Stone (body organ soft 'tissue' stones), B Tended Stone (carrying marks of treatment).

125 pages the whole text..

---Page 220------------

On the Count of

Three

3.
 One of our reasons for a mutual co-operation was to find out in what places, and under what

circumstances we can find similar specimens of these rocks; in different regions and continents. Our goal was to draft an approximate circumstance, and a hypothetical climate, under which soft-tissue material could have been preserved in such a pristine form.

4.

5.

6. The region where Stephan Culbreth lives is a humid, flat-land, ocean-based environment; typical for

western California. His opinion that 'Mystery-Stones' got preserved due to high-salinity and ultimately fossilised is based on his 'Anchovies theory' which uses salt (brine) as a natural preservative. Steven Culbreth claims that: "The climate was very different from the one we know nowadays. Pieces of prey accidentally dropped into hot springs, water, or bogs, and the agent of salinity played the role in preserving soft-tissue."

7.

8.

9. Benjamin Schmidt

0. My data are based on the evidence found in the Czech Republic. So far, I can claim only one locality in

the capital Prague (map A.). The site in question is a send-stone outcrop overlooking a slope; connecting two streets with a differing altitude.

1.

2.The finds in the site comprise:

3.2x hearts

4.1x eye

5.1x flint (prehistoric, Celtic)

6.1x paperweight (Celtic)

7.1x fossil egg (most probably bird`s)

8.1x treated stone

9.

map A.
  Dinosaur eye

125 pages the whole text..

--Page 221------------

On the Count of

Three

0.

1. The debris from a work-ditch exposed a couple of tools from pre-Celtic/Celtic period. A question hangs

over the fact whether the palaeontological finds got mixed up with the ancient tools from an upper layer, or got exposed from the same one.

2.

3. These attributes (as I call such palaeontological phenomena collectively) are usually /if not always found

in groups. Whether they present a sort of nest/roosting place (migratory) /dinosaur habitat/ stereotypical ritualistic behaviour, we don`t have enough evidence for.
 or, a

4.

5.A polemic:

6. The lay-out of the environment/landscape in the mesozoic Czech Republic is as obscure a matter of

discourse as the strata that hide the secret of the floral and faunal situation in this region.

7. One of my polemics draws on Denmark and the famous discovery of the bodies buried in peat-bogs.5 I

think, that one of the possible explanations might lie in the food remnants and the parts of butchered food having been quickly submerged into a soft, boggy environment. The absence of oxygen and the presence of preserving chemicals contained in peat gradually turned anything living a good candidate for a 'bog fossil'. A gradual change in conditions and temperature then played its role. The bogs dried up and the soft-tissue

fossilised.6

Collective Data
 C./S.
 14.2. 2013

8.

9.

5 Wikipedia, The bog Man of Denmark, en.wikipedia.org/wiki/Tollund_ Man 6 See. Benjamin Schmidt, Calcification Fossilization [treatise]

125 pages the whole text..

On the Count of

Three

0.

Calcification Fossilization

1.

2.

3. Calcification is the process in which calcium salts build up in soft tissue, causing it to harden.

Calcifications may be classified on whether there is mineral balance or not, and the location of the calcification.1

4.

5.
 Imagery:

6.

Brass Stone, find no. I. A

7.

8.This theory is based on the finds in a Prague locality (map A.).

9.After a thorough research into this matter, I came to the conclusion that soft tissue, or tended stones

(generally) were preserved according to the following circumstances:

0.

1.Sandstone bedrocks from the Cretaceous Era on map A. might have had the shape of cliffs overhanging a

saline sea. So far, evidence presupposes the presence of large birds /dinosaurs.

1 Wikipedia, Calcification , http://en.wikipedia.org/wiki/Calcification A Brass Stone [Body organ soft (tissue) stone]

125 pages the whole text..

---Page 223------------

On the Count of

Three

2.The cliffs most probably served as a roosting place and a migratory habitat.

3.

4.Chemistry:

5.Calcite can be produced directly from limestone/ also sandstone.

6.Calcium Chloride & Calcium Carbonate precipitate from these rocks through the presence of

Hydrochloride (HCL), or acidic rains.

7.

8.
 Calcification Table:

9.

0.
 - ironisation/ opalisation mostly veins

1.Soft Tissue – Calcium (calcification) – Quarts – Talc
 fat

1. Feldspar

2. Schist (silificated) 3.

muscles

2.The theory of Calcium Chloride & Calcium Carbonate lead me to assume a very cold climate (due to the

CaCI2 & CaCO3 precipitation in low temperatures) 20°F, -30°C..

3.

4.

5. Metastatic calcification:2 deposition of calcium salts in otherwise normal tissue, because of elevated

serum levels of calcium in blood, which can occur because of deranged metabolism as well as increased absorption or decreased excretion of calcium and related minerals, as seen in hyperparathyroidism.

6.

7.

8.

Imagery:

9.

Alpine schist, find no. II.

2   Wikipedia,   Metastatic   Calcification   ,
http://en.wikipedia.org/wiki/Metastatic_calcification

125 pages the whole text..

---Page 224------------

On the Count of

Three

0.

1.II. Further field research proved the presence of Alpine schist, and skipping stones (one most probably

crafted), both associated with glacial presence.

2.

3.

4.

5.

6.

7.

8.Further data:

9.Metastatic calcification can occur widely throughout the body but principally affects the interstitial tissues

of the vasculature, kidneys, lungs, and gastric mucosa. For the latter three, acid secretions or rapid changes in pH levels contribute to the formation of salts.3

0.

1. Imagery:
 4

Brass Stone find no. III.

3   Wikipedia,   Metastatic   Calcification   ,
http://en.wikipedia.org/wiki/Metastatic_calcification   4   Jale   Rosen,
Photo, http://www.flickr.com/people/pulmonary_pathology/

125 pages the whole text..

--Page 225------------

On the Count of

Three

2.

3.

4.

5.

6.

7. Additional imagery:

8.

9.

0.
 Crystallization Temperature - Aqueous Calcium Chloride5

1.

5   Engineered   Solution   Guide,   Figure: 18.,19.,
http://solutionsguide.tetratec.com/index.asp?

Page_ID=734&AQ_Magazine_Date=Current&AQ_Magazine_ID=2238

125 pages the whole text..

---Page 226------------

On the Count of

Three

2.

3.

The crystallization point is determined by cooling a brine until salt crystals form, and recording temperatures at various times during the process. Figure 19 shows a typical cooling curve for a brine. Note the three points along the curve. First Crystal to Appear (FCTA) is the point at which salt crystals first form. The formation of salt crystals generates a small amount of heat, which causes a slight rise in the solution's temperature. This higher temperature corresponds to the true

crystallization
 temperature
 (TCT)
 of
 the
 brine.

Once crystals have formed, the brine can then be heated until all the crystals are redissolved. The point on the curve which corresponds to the temperature at which the salt goes back into solution is labeled Last Crystal to Dissolve (LCTD). As a general rule, the TCT is the most commonly reported crystallization point. In practice, the TCT is extremely valuable since it is a strong reflection of the composition of the brine. It is, in fact, the most reliable and reproducible measure of the

safe
 working
 limits
 of
 heavy
 brines.
 The
 measurement
 of
 TCT
 is
 governed
 by
 an
 API
 protocol.

In the case of multisalt brines, the least soluble component will crystallize at the TCT (Table 50). Thus, if a heavy brine is contaminated with minor amounts of

125 pages the whole text..

--Page 227------------

On the Count of

Three

NaCl or KCl from formation brine or seawater, the TCT may be shifted to a much higher temperature. This is due to the limited solubility of NaCl and KCl in heavy brines. Although the brine at the altered TCT may appear cloudy, it can be cooled to the original TCT with no further crystallization occurring.

4.

5.

6.

7.

8.

9.

0. Map Imagery:

1.

Map A.

125 pages the whole text..

---Page 228------------

On the Count of

Three

125 pages the whole text..

--Page 229------------

On the Count of

Three

2.

Map A. (saline sea)

3.

4.Additional data:

5.

6. Hydrochloric

acid

From
 Wikipedia,
 the
 free
 encyclopedia

Jump
 to:
 navigation,
 search

Hydrochloric

IUPAC
 name

acid

Hydrochloric
 acid

Other
 names
 Muriatic
 acid,
 Spirit
 of
 salt

Identifiers CAS

RTECS

Properties

number number
 [7647-01-0] MW4025000

Molecular
 formula

HCl
in
water
(H2O)

Molar
 mass
 36.46
 g/mol
 (HCl)

Appearance
 Clear
 colorless
 to

light-yellow
 liquid

Melting
 point
 -27.32
 °C
 (247
 K)

38%
 solution.

7.

8. Conclusion:

125 pages the whole text..

---Page 230------------

On the Count of

Three

9.

0. Although we can, to an extent, account for many phenomena, the relatively prolific absence of skeletal

remains (particularly concerning the field research in Granada, Ca., Stephan Culbreth) and other significant matters still allude our understanding.

1.

2.

3.

4.

5.

A find from locality A, map A

6.

7.

8.

9.

0.

1.

2.

3.

4.

5.

6.

7.

8.

9.

0.

1.

2.

3.

4.
 Benjamin Schmidt ,
 Prague, ČR 27.2.2013

5.

125 pages the whole text..

--Page 231------------

On the Count of

Three

6.

7.667.

 …................................

8.

9.

0.

125 pages the whole text..

--Page 232------------

On the Count of

Three

1.

2. Ph.D. Work

Benjamin Allan Schmidt

Věnováno všem, které jsem potkal. Ať pochopí, že toto je jen žert.

Dedicated to all I met, may they understand that this is but a joke.

[dosens of compiled esseys of students from Italy, Cyech rep. Ireland, France]

Ben Smith

Spitfire from Eire, on front page

On The count of three slam/sane peat boge Honeymoon Ben Smith Susi sister

spanish, italien, french german, gealige, chinesse, portugees

her is Bobo,

Martha, Merha

shakuhaci Ksindel make up names / we name Eir ame friends..

I claim that the use of profane language stems from uneducation, and I call this disorder

Benjamin afer my name.

125 pages the whole text..

---Page 233------------

On the Count of

Three

Foreword

This Antropological text-book is designed firstly for my Ph.D. purposes, and may not find its reseller due to its contents. I nevertheless claim that the book was designed as Charles Bukowsky Pulp and is aimed for adult people who are interested in ever the expanding horizon of this science. I lead you through several stories of a congnonimous scrutiny of people hassling for work. Their problems with education, religious codexes and mutual friendship - closely bordering on social and racial unfriendliness (hatred). Hadda love your fellowman might be an another title how to express this book and its textual content. I lived in the Prague's expat. communities since my early age and spent 15 years among them in several countries. 3 years on Erasms in Italy, Ireland and France where I was occassionaly obliged to eat out of dusbins to make ends meet.

Curriculum Vitae Benjamin Schmidt

Dosa ž ené vzd ě lání:

1999 - 2003 St ř ední um ě lecká škola Designu, Praha.

Staré grafické techniky (suchá jehla, m ě diryt, litografie, aquatinta, lepty).

Po č í ta č ové programy: Photoshop, Illustrator, Maya, 3D Studio Max.

Kompozice písma, Typografie.

2003- 2004 Ro č ní studium Angli č tiny, The Language House, Praha.

125 pages the whole text..

--Page 234------------

On the Count of

Three

Ameri č tí lekto ř i -Zakon č eno diplomem.

2007 – 2009 Studium na Filosofické Fakult ě , Karlova Univerzita, Praha.

Obor: Anglistika-Amerikanistika. (studium Angli č tiny, Irštiny, Hindštiny a teorie anglické a irské literatury a dramatu).

Studium ukon č eno z rodinných d ů vod ů po 5. semestru.

Zam ě stnání:

Archeologický asistent
 (2004)

Spole č nost archeolog ů ARCHAIA

Archeologické výkopy pro dnešní Paladium (Rudolf II.), Veleslavín

Dale. Centrum2/5, germanska kultura, Podbaba,2/5 mes(Paleolit). Zlicin 5 mes geologicke anomalie

P ř ekladatel
 (2005- a sou č asnost)

P ř eklady environmentálních text ů - archeologie, geologie, architektura, um ě ní. spolupráce s australskými a kanadskými programátory na tvorb ě webových aplikací.

U č itel Angli č tiny druhého stupn ě ZŠ Pet ř iny-Sever

1. Výuka Angli č tiny podle kurikula.

(2006-2007)

2.
 Ka ž dodenní p ř í prava na hodiny (1 a 1/2 roku zam ě stnaný na nad ů vazek).

Datum narození: 19.12.1983

Od roku 2011 se zabívám teorii p ř í stupu k um ě ní.

Aoutor nekolika povidek a jedne knihy On The Count of Three, The Mitzubichi.

Tools as a means for self regeneration.
 http://gallerymarvels.com

Lives partly abroad.

Ve své tvorb ě se soust ř edím na:

(P ř í stupy k výrazovým prost ř edk ů m, Um ě ní, p ř í stup a seberegenerace).

Vystavoval jsem v Divadle komedie jeste za Lucie Vondrackove , pote
kousek od betlemske kaple , Klub

125 pages the whole text..

--Page 235------------

On the Count of

Three

Kastan Praha 6 (2005), Majk L ' Atmosphere Praha 6 (2013). Chystam
nyni vetsi vystavu s koncertem v Dobre Trafice na Ujezde .

Znam se osobne s Vladimirem Mertou, Vladislavem Matouskem, Oldrichem
Janotou a Vohtechem a Irenou Havlovimi. Jsem dobrym prateli s Jirim
Dohnalem, Petrem Korbelarem, Ondrejem Smejkalem, ci Radanou Lancovou.

Snazim se propagovat umeni , ktere ma co rici zkusenemu vytvarniku , i
laiku , metodikovy , I exuberantnimu hromotlukovi . Jsem byvaly pedagog ,
a zabivam se vlastnim dogmatem nastroje jako prostredku k seberealizaci .

Předmluva

Tuto knihu /Ph.D. praci jsem psal se třemi body na paměti. Kniha by měla
byt strukturovan8 jako u4ebnice Antropologie, měla by být co nejúplnější
a měla by skončit. O tomto oboru se někdy nepíše úplne snadno. Je to však
snazši, než literární pokusy ze záhrobí. Nicméně se cítím povinen
justifikovat hlavně bod druhý a tím tedy onu úplnost knihy a její podobu.

125 pages the whole text..

---Page 236------------

On the Count of

Three

Pokud vyrůstáte v bilingvním prostředí a druhý jazyk se vám v útlém věku stane mateřštinou, vypozoroval jsem u bilingvnich lidi tendence, kterých jsem si dlouho nebyl jist. Jde totiž o jistou rezervovanost a jakýsi žoviální přístup k jazyku a řeči jako takovým. Kniha je tedy kompletací mých myšlenek 15ti let v České Republice jako cizince a tři roky na frontě v Éře. Flyghts over Great Brighton under heavy gun flack Sister Susanne my girlfriend.

Tam, kde nám chybí slovo, nás často napadne jiné v jiném jazyce, ale mám zejména na mysli tendence přicházet s novotvary a absurdy. Nicméně mi přišlo nemoudré psát knihu s až tak velkorysým přístupem. Proto jsem se rozhodl využít dnešní zažité žánry a namíchat tak koktejl. Cosi, co by se dalo přirovnat k mutaci volného psaní a slam-poetry a dát tak knize jistou volnost. Takovou, kterou mnohé dnešní literární žánry a učebnice spíše postrádají. A navíc tedy, co se té vólnosti týče, vyjádřit zde jakousi vólnost duše, spíše než onu ne-vólnost materiální svobody, v které se dnes tolik 1 pohybujeme. Kniha je rozdělena do tří částí. Jsou to tři kroky k poznání knihy. Tři etapy poznání v retrospektivě osobnosti - aneb najdi chybu tam, kde mi to neseplo - Arteterapie duše.

Malaric syndromes in a free-verse prose.

I call the use of swearwords and profane languages a possible disorder steming from a difficult Family life,disrupted family or overeducation, or education without a good positive feedback or illustration into life. Preparation for adulthood under positive circumstances with possitive and healthy attitude towards the mentioned individual. That he/she grows up in positive circumstances, happiness, illustration and instruction.

Benjamin Allan Schmidt

Syndrome/disorder:
 BEN JAMIN

11 1. follow with a pen to check pages (rude and hilarious) be apsent of rudness in a belingual state of supreme equilibrim. And tighten the screws in both languages.

125 pages the whole text..

--Page 237------------

On the Count of

Three

ABRIDGED

I.

1a.Footnotes/ The weight, and size differences extrapolation was very
fashionable in the overreach of the 21 st cent. Food, and the lack of
food is a very modern thesis bolonging according to me to almost
nostalgic Kavannag, Ire. Or, posmodern Art Nouvau reccurence in the 21 st
cent. I Stuffed Zebra/ Tender is the Night / Dick Frencis, Zoo - drawing
of animals science and education, desert for palaeontology and Palaeo-
botany, zoology, the Prague Zoo, Zoo in Dubhlin, and in the South France
and Italy.

Fast camera in films, and flexography can maintain its iffect on a
cooling space reaccurance, just as in modern photographs and the space in
between. Large, diametrical differences ar known only inside fields, but
never overreaching them.

2h/Footnotes - Continuation several pages After. Sexist language above.

A short userpt on Trob

125 pages the whole text..

--Page 238------------

On the Count of

Three

Film competition I.

The film is based on an idea of education, elation and jazz or R&B production. I'm singing west Virginia

pride in a paradigm - those are mormons who travel to Africa to educate and spread belief.

I'm coprolitic in my expression / Book 1. as well, - and I claim that I'm trying to use swearwords and demuring all-recital language- to solve racial problems & fluency.

. .

Famil
.
Enfant

. .

• The film is devided into ½, 2/3 segments (famil, enfant / family, children..

Melodramatic, Allen Petterson opens up

The film is based on minimalism, almost Dada, + new aspects for education / science

Structure of the Narrative

Me - He

You- She your partner / the amalgamation, stilization of a Narrative language is a matter of polemic.

In true deep therapy. Ends up in Aristotel's cave. Normal Oidipus.

Two gods, two religions, two directions of submittion.

After Shock feelings (You constantly search for an excuse, omission, or an excuse that it's going to be better. That feelings of danger will be left behind. People are willing to resort to incredible ends to reach peace.

You can cherish bad feelings on the bases of your language inedequacy. Things feelings should be named,

pinpointed so that they are understood.

I tried to borrow scissors - I was several times told that it's impossible. I appologised.

Pertaning mindwork

· that you eschew from eating something and you remember the past experiences. -

During deep therapy and before after, - you should be warned that you should eat such that yu are only

happy -at the same thime ypur behaviour has to be, or should correspond with mundane.

· I here think .- I'm interested in behavioral arythms / enclusively and collectively called rythms -

possitive (inclusive exclusive negative).

I undergone a fast and in chill I recorded a song that was supposed to be in geanra #rock

125 pages the whole text..

---Page 239------------

On the Count of

Three

more affinite to shanson, classical in Aretha Franklin

I hereby disclammer all possible notions and evocations that I use any inapropriate versing (and hereby would call any such presumptions inapropriate, absurd and prefabricated. I'm solving if surroundings and instruments do tend to teach you almost manners.

I almost claim that harmonies -. and proper harmony stearing can lead to a betterment of verbal or

{speech functions. That's why we understand why education is so important.

I would be almost willing to theorise that a musician whu atteins a very good learning of ragas; {upanishadas, and mastery panatipata suta..

can better his/her speech (automaticaly, in succession), we is three, food. 1.Study of how to attein an

instrument

Theory.?/ People nowadays explicate on behavioral a-rythms with the possibility of reaching spiritual climax.

I claim these aspects impossible and call them illusional [disilusion, on e of the roots of evil / Buddhism. ---------------------------------

I tried several times to express profanities, that lead me to understanding that I was in my fast cold and shaking. People with belinguality have much greater perceptive and receptive knowledge

Behavioral extensive rythms [positive, normal, negative and based on general good education, religious codexes of speech, behaviour and job attandance. Person should be understood that he / she can live in a mutual respect /according to religious codexes, good manners, education, and the prospect of atteining a payed job, with + atteining at least one instrument.

Problems arising from inundation

Thesys underline, logical prognoses in afternotes:

the problems ensuing from this as possible behavioral changess, understanding of certain patterns of speech and vocabulary insuing in a pattern malady / pinpointed logical aspects, bases for educational, religion codexes behaviour, upbringing and growth of a man (individual).

I here claim that suck a cock an get a horse! (breton language) That my mind should blow! Is a logical faximily of a prelude, a hyposentence for a tonguetwister, new conage of a fraze - Ensuing further reel on music and education. I claim that coprolaly (unhindred utterance of coprolaly, can better your speech patterns and make order in your vocabulary, fluent your speech and logical aspects of it. ----------------------

This suden ubruption of a mood I call Benjamin, Parents ever growing interest in the life of their adult children. Benjamin Suck a cock and get a Horse! Do you like it, I dont like yhe way you hessed up the toaster. Interupted sentences, no links between a fluent logical flow of a language as crux, and the flow of a conversation + normal obsedance of an exhausted conversation with happines flow.

125 pages the whole text..

---Page 240------------

On the Count of

Three

This simily of a cock sentence (a trigger), a parallel to a family
life/rebuff on a logical sentence of your parent. Abundance of possitive
thinking, destruction of dreams, and links on a logical feel, and rhythm,
understanding of a day. Exasperation, ensuing frustration and envy. -
Further/ logical peacing together of logical aspects connected to moods,
images, and feelings.

Doesn#t have a day. A whole day feeling through, logical prognoses on a
walk out, visit, laughter and normal feeling and channeling happiness,
without hindrance or logical prognoses.

Famil / Contrary to that - Life in a Family of parents. Felling safety,
peacing together of logical prognoses,

its often shatter, wave line _/-_-_-_-_- , _____-___

I claim that the frustration of current young people 30,32 years old for
independent family life is a relatively palpable and strong aspect of the
current society, and sociological circles. Stance 1. Appandages:

Filmography is a powerful tool. We have to understand that Caroux and
Donstena my plays. Troubles, empty shipyards, 2. Musicology. Jane Awsten.
Jane Godall, Jane Srtre..

Problematics of intertwining of mutual prefabricates in mutual
correlation..

There is an interesting barier to solve that we would entwine spiritual
instruments in music such, that the ether would be visible and would
widen to a visible and palpable scope. A similar problematic might arisen
in a credit card being million, or billion by minus. There is a strong
simily to the current trend of feshion and the possible extrabold on
logical live norms. We've always aspired to the ideal of religiousness
and happiness, and this notion should always be for mostly exaggerated
and put forward for the practice of populi. The problems with everreach
into different societies is an automatic flow. We should always behave in
the best respect and understanding without harme of the others, in
education, understanding and the possibility of a happy life. You should
never result to any ignominy for any disillusioned reason as such act is
in a contradictory understanding of your framework and structure.

A logical life, religion of love and mutual respect with procrastiny in a
relative or absolut happiness without harm.

Pupei

Puppy, Pie, Pee, Pu, Pay, Pe, Mir Loutka {low, tetiva, ka, you need to go
to the toilet when you see a carver. Industry, Uistacht, Usine. Vin Vine,
Lumir in Mucha, Lublan. Medow, Poplar + stag,cow, herds/ craft & agricul.

125 pages the whole text..

---Page 241------------

On the Count of

Three

I can explicate almost any word in several langues in several langues by heart.

I have an inclination to write several books: on nature.. Antropological thesis.

1. /do . A word meaning incessant empathy, love and possitive thinking without harm. Act 5, everything that is pronounced, more unpronounced and means an utter bliss, without the harm of others or any living thing.

1d/Footnotes - Possible parallel and a simily to Domini Derry, my friend. There is a powerful link to the Prague, Czech Pen Club, Arnost Lustig, Viveg, Where the Dog is Burried. The below text almost like The Eternal Lightness of Being.

2. Footnotes / I aspire to the idea that Antropology, modern anthropology as a branch of science can be explained in a textual form. Ulysess, by James Joyce is a wonderful example of an anthropological, sociological understanding of a nation, nomenclature.

I deem this above text is a good stylistic of a Dada textual form, verging on a free verse. I'm not sure if the possibility of the use, and utilisation of swearwords is even possible in current new literature. I still claim that due to the change in current critique, for a better industry and the look back on the, let's say the literatre of 80s, 90s of the 20 th century, Czech Repub. Might lead to a better view, and understanding of the oncoming ways of writing and modern literature (Postmodern literature in my view. Basically a new beginning like Art Noveau, Art Deco 2035-45.)

Arhhhhhhhhhhhhhhhhhhhhhhhhhhh. Horssssssssssssssss. Look at me now.. Juju, Pupi, I have 'cha, smeck ma, I have cha, got cha on cha, litricha, kari kari. Toor, like rue.. France Uistacht… Gealge na Loabhar sa rue..

3.Footnotes / The possible extrication of onomatopeia, exchage of word order. Can even lead to new stilistics. (Gealige, Breton, English, French…)

Exchange of word order. Border (bordering on something new) like Labrodie, diew..

…… Look, lok.. I have cha, look at the tv. Po, I've extricated this fayon from under my chest od clothes. Have cha like a palm pad..

Box of match, La brike… Stupido.. Limet…. Like Sherbet lemon..

Lemon grass, Rai grass like Iane Andersoneeen…. Whooo, ye Darrrh. Ana cha, anoorie, Toyo……. Like Tat… Urpiiidiii… I have never seen some apples being called sherbet. Pommess, [pomes] pomes du terra, almost like

125 pages the whole text..

--Page 242------------

On the Count of

Three

4. Footnotes/ We deem this last textual form an unique ensight into the forming of words, it's changed structure and altered meaning.

I remember my second visit to Italy living hood. I understood a lot of my Erasm in this respect. Like mausam he, that mind changes with a scope of recognition of a foreign mirrored by the local interaction.

5. Footnotes/ The eirport to Bergamo, just as the Roma Fumicino was like two roads into consciousness of what Italy really represents.

I arrive and shake hands. The Santacittarara Monastery is woaven in mystery. Ajhan Otala, (Ashin) is a

senior monk and the highest representative of this place. We talk a lot about everything (2).

(2) The short history of almost Everything, just as the Hitchhikers Guide come to mind.

I'm willing to meditate.

6.Footnotes/ Digest what I am. An Antropological thesis.. A Frase never seen. A possible existence in future books, and literature of fiction and interest geanras. A joke. (Metafiction).

I meditate in the morning, I eat once a day before noon.

7. Footnotes/Not even slang. Expressions mediate to corelate with reality and the illusionary narration.

I understand the late night meditation is on a verge of happening. I go to meditate. Full of ripe judgement I meditate.

8. Footnotes/ rascals. A possibility of omitin the word would lead to disinterest and not understanding the sent. In a laughter. Like incandescent children.

Foxy.. I have you.. Tri. Bit by bit, number extrapolates..

9.Footnotes/ over passing in logical judgementation. buddy is…

1b/Footnotes /It's interesting if you as a critique would take this sentencuation immediately as dearing. You have to understand that Palaeontology, Botany, Anthropology / the branches of science generally are only just for a few decades being religiously extrapolated. That we only in recent books find these parenthesis on religion, exploitative on religious understanding of scientific possibility and religious

125 pages the whole text..

--Page 243------------

On the Count of

Three

demurement.

interested in the rudiments of Paleontology.

"Look Darling," he said, searching for an excuse.

1C/ I spent half a year on an Ire Beach, Quartenon 2547. Surrounded by seals, giant

walruses, sharks and dolphins + whales.

Wesk Cork, Co. Kerry, Dingle - Allan P. Natural Reserve. Dingle, Kerry. Is claimed to be one of the most preserved piece of land in Ire.

The forces there and the local habitat corresponds almost to high tides, over heatching

of small species, and the possibility of high winds, warm wint..

"I have a lecture at ten. Speak to me about my work tomorrow at 10:00... 1a/Footnotes -

gibson lespoul, gypsy, gypsy King [musician], giro, 2, ro, Roma, east, Po, PE, PIE, Pey /Bread, Peace, dark, cake, spin..

There is a possibility of this word altering its meaning for a completely different thing

in the 21 st cen.

can barely speak Engli but prove to be quite a miracle. Have a good day."

2/ Here I ostentatively give a notion that I might be from a gipsy family. Originaly possibly from south America, India. Bhenimin Allan McGobhain. The one who holds a rock of prestine beuty and can be its maker. Tolkien /

Aldus Petticoat was an old man. He held a Ph.D. in Astrophysics, Europeen

3a/Footnotes - Europeen Almost like the Ulysess by James Joyce [Ireland Gealtacht, Lunecy.. Possible overlived express. From FF,Uk. Newver heared of in Ireland in a daly conversation/

Linguistics and a completely new science dealing with predicting your own death by

the luminosity of morning suns…

3b/Footnotes - The winter and summer equinox. Celtic hollidays, and celebrations. Mount. Tara, Giant Causeway, Chusulain May, Sleav League.. All visited by me. Benjamin Schmidt [Belingual \children education, growth and mutual corespondance in groups and the society, company, living hood, work a Job maintanace and monay intake, prosperous life in happiness and harmony according to the highest trends.

Trends - Low[e] intrication, entwining into the group, family and the society by undermining

125 pages the whole text..

---Page 244------------

On the Count of

Three

influences. From within by missunderstanding of logical trends, from
without by forcing an individual into a different scope of life than
which corresponds with his/her its individual talents, logical scope and
understanding [inprint] /Melevolent influences, ensuing disagreeability,
unhappiness.

Normal A relatively standard way of life + with a relatively logical
presumtions,

understand ability, possibilities, ups, and downs, and logical feeling
boundaries. [I'm a very strong seeker of a Buddhist notion. There is a
lot of very strongly religious people in Lunetic Asylums and among the
impoverished children. I claim that the bases of this notion and its
groundwork can only just have one offshoot for a possible disease. Malady
[minor problems usualy in cognisance and plus with born in disabilities.
The more powerful is a problem of the inability of the social circles and
the surrounding peer group to give an individual enough knowledge for its
religious ardency, its vents, logical extrapolaty and understanding +
respect [for an individual].

Highest a possible large money intake people, nonsensical performances on
the person from the side of the peer group. Threatening, blackmailing.
The rest like Normal.

"So, I'll make you some pasta-!"

1d/Footnotes- I here in the last couple of sentences exchange rude and
profane expressions [that might sound hilarious, Charles Bukowsky, Pulp,
Even Joseph Heller [The Porn, or Trainspotting by Irvin Welsh is a
completely different geanra for canversation or literally endeavours],
even a word Nice can sound rude or obscene in repetition or different
wordings, phrasings or locutions].

1e /Footnotes - Fi Fe Food - is very Fren.. Mauze, like Mausam he,
Hindi.. Jidlo her feed dloub, wallop

or a spoon. A man feeds a woman or a child.

Foo, foo, foood…

and in his mind, thoughts of incoming danger

were ever present.

11/a Footnotes- Aldus Petticoat suffered from a chaseing

desease, or prsecution as an important man. He was a disillusioned
individual maladed by years of teaching and study. I would say that a
similar case is possible to be found in real life, a problem of
overeducation without enough justification. I call in Charles desease.
Without any logical background of affinity.

of course, what comes after it.

12/a Footnotes- A tipical disorder now often found among middle age people and singles, who feel they have evergroun the possibility of having childen Husband/or a Wife, an independent family life. Truly existing and arising in people who are just in a phase of would like to obtain independence in a new relationship. They often

125 pages the whole text..

--Page 245------------

On the Count of

Three

break down befpre finding peace in their loneliness, or reclusiveness [seclusion /Film 2, Couroux].

preferable. And now, the last five pages were

written in shorthand.

13a/Footnotes- A tipical problem with concentration of an enough duration not to alter your plans for the longness of an essay, holiday, a normal day activity, problem with laptop ediction [fluctuation syndrome, computer take in].

Fluctuation /walking aimlessly for a logical goal knowing that its just a lure or a stratagem to maintain, retain a logical sanity of the day, subconsciously knowing its deeper purposlesness, ensuing logical superfluity of days with a very high happiness stigma feeling. I call this syndrom Evelin. I do not think that the treatment of this desease or syndrome should reach beyond the scope of accepting parents and educating the whole family of the importance of the individual suffering from this diorder. His/her family should be informed and educated about the needs, scopes and views of this individual without the individual's interaction with them. Putting the individual into the learned family which should not further ostracise him/her with nonsensical questions about education, spiritual codexes, or belief. Such individuals suffering from this disorder should be received gifts, a notion that they might change their names, get christened without their parents presence etc. The religiousness, and spiritual codexes of such individuals should never be doubted, they should be asked questions hinted towards their religions, education, and should be admired for the knowlidgebility of their answers. They should be almost helped in obtaining a partner rather then just being instructed how. The syndrom is most often palpable among adults of 28-32 years old but can arise in even younger people or in late age, anytime a person looses his/her partner bond.

125 pages the whole text..

---Page 246------------

On the Count of

Three

"Hey." Look, how she swims in the pond. ""Perhaps, I should write a long paragraph."

"Since I was a little child,

A mouse [topo]…" "Yes, yes…"

5/uFootnotes- Slug - similar to snail, Couroux Film festival

--

Employees themselves did not go to such

far reaches..

6/Footnotes - Cig /a/ret gipsy, black and white, cini hindi/yellow {and white] Couroux film comp.

anymore. "Hey, Jardeen, let`s go for a

trip."

14a/Footnotes- Antropology is a science very important for a parallel growth of sociology and has a strong impact on a lot of branches of science. By the study of bones, we can lear a lot about maladies and thereout a lot about the past life, manners, ways of living, we can simulate and mimick old life patterns for a historical efection.

to tear trousers!"

14/b Footnotes- You would be surprised that this insult can in some countries be one of the worst insults you can ever say, and therefore you should never use it. It is logical but not often understood that we can buy a swearword, slang dictionary to see and understand certain sentesuation, wordings or slang,. You should be nevertheless aware that the use of profane language in publick and most geanras is almost forbidden due to a common sense and international rules of writing.

"Furlow!" Said the scientist. And whistled.

15/Footnotes- From the past History in many countries around the world we've

125 pages the whole text..

--Page 247------------

On the Count of

Three

gleaned the information that some people are willing to ostrecise others for being more happy than some other people.

For example in the Cyech rep. it was

called Iron curtain.

I was given a rise and a new position,

I stood in the middle of the room..

20/Footnotes- Suicide can be a terrible way how to get out of trouble. Depressions in early age should be treated by a very possitive company and surroundings. Unhappy replationships, unstable family life and debts can lead to such depressions.

his study was great, look!

"The food is wonderful, Darling!"

10Footnotes/ the problem of unfinishing words and sentenciation is

similar as in Flour de Mal. The cursed poets are a very interesting and prognostic picture of Pari life where you might suppose that when you travel really far [expatriotic problematic., job attenment, a very good prosperous living hood. There may be a possibility of completely changing the way of life of an individual. Mind broadening and a recursive comeback to basic inner values and understanding. Happiness. A possible attainment of hobboes and children dreams which might have been up to then completely overwhelmed by

125 pages the whole text..

--Page 248------------

290

On the Count of

Three

surface prognoses.

Such child dreams and hobbies are very important to be further developed and helped by others in developing as they represent the most crucial bases and founding stones of a personality and its growth/ A possible problems in modern society may go against these intrinsic values and therefore lead to unhappiness of the patient/individual and its core is represented by the dissilusionment of the sister/Doctor or

practitioner. + possibly trying to find analogy in Parents or even thinking that you are similar to your parent when you are usualy similar to your grandparent of which the Psychologist does not have any information. He/she deems them over excess. Simplification in current science can lead to a mal work. Enough prognostic evidence can lead you to the same conclusions as at the beginning. The current psychiatry and psychology holds the notion and works on the bases of altering you logical rutes and ways, and trying to change the ways and desrupt your logical structures due to an accident. I think there shoud be a mediator, medium way, middle path to go exponetialy from the last two segments askance the way up.

.

. Where psay. Should be at least aimed to..

.

Understanding [comprehension of a ripe personality as even the bases for a session [first intake

of a patient]. I'm an ardent seeker of this notion of practice. Desruption [disrupting basically elements and structures]

7a. Footnotes/ Pictures, happy illustrations and caricatures aim to signify what is hilarious and easyfied. Lofty in a hilarious manner, Laughter-trigging.

125 pages the whole text..

---Page 249------------

On the Count of

Three

II.

Jenifer got up and passed me the salt.

2aFootnotes/Problems with belief, understanding of religion a political structures according to current trends and notions lead us to a notion, Laughter. A possible analogy and simily to a complete corruption of reality is in a man like Kavannah breaking a broom or a rake, shovel and walking away. Then we understand the logical leadout of things, ways. We should still understand that the book On the Count of Three Slam Ben Allen is about Ire. Quartenonb5267 and their religious codexes problematic and, ontake. The problem with learning, education and logical system changes. The result is a try for finding of a new common sense, analysis, and a new viewpoint on the cultural undergrowth. That is, a higher position.

and my children too scared.

6/Footnotes- It's interesting to suppose that the story is solving current problematic of religious biases, and a look upon religion in general. I would claim that there is a possible offshoot of the story in a comparison to Israel and its good living hood. The problems of a family life and the upbringing of children.

"I want ham, Mum." Complained

125 pages the whole text..

--Page 250------------

292

On the Count of

Three

Ben Smith

Born in 1983 in Prague, Benjamin Schmidt studies a private school of Art in Čakovice, on the Capital's outskirts, after which he successfully enrolls for the FF,UK, Prague. An artist, folk-singer and translator; his first book offers a hillarious insight into the human life, and condition. Three other books on. And sull

Gallerymarvel s.com

3.

4.

5.

6.

7.

8.

9.

0.

1.

2.

3.

4.

5.

6.

7.

8.

9.

0.

1.

2.

3. ....

125 pages the whole text..

On the Count of

Three

4.

5.

6. Discovery by: Stephan Culbreth

7.

8. Calcification Fossilization theory by: Benjamin Schmidt

9.

0.

Owen R. (1855), Lectures on the Comparative anatomy and Physiology of Ivertebrate animals, Delivered at the Royal collage of Surgeons . Longman, London, 689pp.

Dawson, J.W. (1888) The Geological History of plants . Kegan Paul, Trech & Co., London

-and Callow R.H.T. (2007). Changes in the patterns of Phosphatic preservation across the Protozoic- Cambrian transition Memoirs of the Asociation of the Australasian Palaeontologists .

125 pages the whole text..

---Page 252------------

On the Count of

Three

-1

Ph.D. Work

Benjamin Allan Schmidt

Věnováno všem, které jsem potkal. Ať pochopí, že toto je jen žert.

Dedicated to all I met, may they understand that this is but a joke.

[dosens of compiled esseys of students from Italy, Cyech rep. Ireland, France]

Ben Smith

Spitfire from Eire, on front page

On The count of three slam/sane peat boge Honeymoon Ben Smith Susi sister

spanish, italien, french german, gealige, chinesse, portugees

her is Bobo,

Martha, Merha

shakuhaci Ksindel make up names / we name Eir ame friends..

I claim that the use of profane language stems from uneducation, and I call this disorder

Benjamin afer my name.

125 pages the whole text..

--Page 253------------

On the Count of

Three

Foreword

This Antropological text-book is designed firstly for my Ph.D. purposes, and may not find its reseller due to its contents. I nevertheless claim that the book was designed as Charles Bukowsky Pulp and is aimed for adult people who are interested in ever the expanding horizon of this science. I lead you through several stories of a congnonimous scrutiny of people hassling for work. Their problems with education, religious codexes and mutual friendship - closely bordering on social and racial unfriendliness (hatred). Hadda love your fellowman might be an another title how to express this book and its textual content. I lived in the Prague's expat. communities since my early age and spent 15 years among them in several countries. 3 years on Erasms in Italy, Ireland and France where I was occassionaly obliged to eat in cheap mlaces to make ends meet.

Curriculum Vitae Benjamin Schmidt

Dosa ž ené vzd ě lání:

1999 - 2003 St ř ední um ě lecká škola Designu, Praha.

Staré grafické techniky (suchá jehla, m ě diryt, litografie, aquatinta, lepty).

Po č í ta č ové programy: Photoshop, Illustrator, Maya, 3D Studio Max.

Kompozice písma, Typografie.

125 pages the whole text..

---Page 254------------

On the Count of

Three

2003- 2004 Ro č ní studium Angli č tiny, The Language House, Praha.

Ameri č tí lekto ř i -Zakon č eno diplomem.

2007 - 2009 Studium na Filosofické Fakult ě , Karlova Univerzita, Praha.

Obor: Anglistika-Amerikanistika. (studium Angli č tiny, Irštiny, Hindštiny a teorie anglické a irské literatury a dramatu).

Studium ukon č eno z rodinných d ů vod ů po 5. semestru.

Zam ě stnání:

Archeologický asistent
 (2004)

Spole č nost archeolog ů ARCHAIA

Archeologické výkopy pro dnešní Paladium (Rudolf II.), Veleslavín

Dale. Centrum2/5, germanska kultura, Podbaba,2/5 mes(Paleolit). Zlicin 5 mes geologicke anomalie

P ř ekladatel
 (2005- a sou č asnost)

P ř eklady environmentálních text ů - archeologie, geologie, architektura, um ě ní. spolupráce s australskými a kanadskými programátory na tvorb ě webových aplikací.

U č itel Angli č tiny druhého stupn ě ZŠ Pet ř iny-Sever

1. Výuka Angli č tiny podle kurikula.

(2006-2007)

2.
 Ka ž dodenní p ř í prava na hodiny (1 a 1/2 roku zam ě stnaný na nad ů vazek).

Datum narození: 19.12.1983

Od roku 2011 se zabívám teorií p ř í stupu k um ě ní.

Aoutor nekolika povidek a jedne knihy On The Count of Three, The Mitzubichi.

Tools as a means for self regeneration.
 http://gallerymarvels.com

Lives partly abroad.

Ve své tvorb ě se soust ř edím na:

(P ř í stupy k výrazovým prost ř edk ů m, Um ě ní, p ř í stup a
seberegenerace).

125 pages the whole text..

--Page 255------------

On the Count of

Three

Vystavoval jsem v Divadle komedie jeste za Lucie Vondrackove , pote
kousek od betlemske kaple , Klub Kastan Praha 6 (2005), Majk L '
Atmosphere Praha 6 (2013). Chystam nyni vetsi vystavu s koncertem v
Dobre Trafice na Ujezde .

Znam se osobne s Vladimirem Mertou, Vladislavem Matouskem, Oldrichem
Janotou a Vohtechem a Irenou Havlovimi. Jsem dobrym prateli s Jirim
Dohnalem, Petrem Korbelarem, Ondrejem Smejkalem, ci Radanou Lancovou.

Snazim se propagovat umeni , ktere ma co rici zkusenemu vytvarniku , i
laiku , metodikovy , I exuberantnimu hromotlukovi . Jsem byvaly pedagog ,
a zabivam se vlastnim dogmatem nastroje jako prostredku k seberealizaci .

Předmluva

Tuto knihu /Ph.D. praci jsem psal se třemi body na paměti. Kniha by měla
byt strukturovan8 jako u4ebnice Antropologie, měla by být co nejúplnější
a měla by skončit. O tomto oboru se někdy nepíše úplne snadno. Je to však
snazší, než literární pokusy ze záhrobí.

125 pages the whole text..

--Page 256------------

On the Count of

Three

Nicméně se cítím povinen justifikovat hlavně bod druhý a tím tedy onu úplnost knihy a její podobu.

Pokud vyrůstáte v bilingvním prostředí a druhý jazyk se vám v útlém věku stane mateřštinou, vypozoroval jsem u bilingvnich lidi tendence, kterých jsem si dlouho nebyl jist. Jde totiž o jistou rezervovanost a jakýsi žoviální přístup k jazyku a řeči jako takovým. Kniha je tedy kompletací mých myšlenek 15ti let v České Republice jako cizince a tři roky na frontě v Ére. Flyghts over Great Brighton under heavy gun flack Sister Susanne my girlfriend.

Tam, kde nám chybí slovo, nás často napadne jiné v jiném jazyce, ale mám zejména na mysli tendence přicházet s novotvary a absurdy. Nicméně mi přišlo nemoudré psát knihu s až tak velkorysým přístupem. Proto jsem se rozhodl využít dnešní zažité žánry a namíchat tak koktejl. Cosi, co by se dalo přirovnat k mutaci volného psaní a slam-poetry a dát tak knize jistou volnost. Takovou, kterou mnohé dnešní literární žánry a učebnice spíše postrádají. A navíc tedy, co se té vólnosti týče, vyjádřit zde jakousi vólnost duše, spíše než onu ne-vólnost materiální svobody, v které se dnes tolik 1 pohybujeme. Kniha je rozdělena do tří části. Jsou to tři kroky k poznání knihy. Tři etapy poznání v retrospektivě osobnosti – aneb najdi chybu tam, kde mi to neseplo - Arteterapie duše.

Malaric syndromes in a free-verse prose.

I call the use of swearwords and profane languages a possible disorder steming form a difficult Family life,disrupted family or overeducation, or education without a good positive feedback or illustration into life. Preparation for adulthood under positive circumstances with possitive and healthy attitude towards the mentioned individual. That he/she grows up in positive circumstances, happiness, illustration and instruction.

Benjamin Allan Schmidt

11 1. follow with a pen to check pages (rude and hilarious) be apsent of rudness in a belingual state of supreme equilibrim. And tighten the screws in both languages.

125 pages the whole text..

---Page 257------------

On the Count of

Three

Syndrome/disorder:
 BEN JAMIN

ABRIDGED

I.

Jordan Bruno was hungry. As far as he remembered, he was always hungry.
It had been six hours since he crossed the border to Sierra Tralala.
"Look at that dust everywhere, Molly." He said to a stuffed zebra
dangling on a piece of string in front of his nose. Half an hour and two
minutes later he parked his truck carrying a cargo of metal rivets

at a gas station by the main road. The sun was just rising.

Bruno jumped out of his cabin, his worn sport shoes tasting solid ground
after an all night drive. That is, if he did not count the short break
close to midnight, when he was forced out of his driver`s seat to pee by
the moon. Bruno wasn`t a character out of a novel. He was, after all,
just a truck driver. His fat body seemed to move in unusual places as he
scurried over the empty-for-miles road to reach an already comfortable
shade of the corrugated-iron roof of the gas station. "Fries and a
burger." He announced to the keen- eyed man, who stood unmoving in the
doorway of the station as if he waited for this sentence all his life.

Hungry truck driver Jordan Bruno bit into his first meal of the day and
wept.

125 pages the whole text..

---Page 258------------

On the Count of

Three

1a.Footnotes/ The weight, and size differences extrapolation was very fashionable in the overreach of the 21 st cent. Food, and the lack of food is a very modern thesis bolonging according to me to almost nostalgic Kavannag, Ire. Or, posmodern Art Nouvau reccurence in the 21 st cent. I Stuffed Zebra/ Tender is the Night / Dick Frencis, Zoo - drawing of animals science and education, desert for palaeontology and Palaeo-botany, zoology, the Prague Zoo, Zoo in Dubhlin, and in the South France and Italy.

Fast camera in films, and flexography can maintain its iffect on a cooling space reaccurance, just as in modern photographs and the space in between. Large, diametrical differences ar known only inside fields, but never overreaching them.

While a completely unimportant man, whom Jordan Bruno certainly was, had been shedding buckets of appreciative tears over a piece of beef all alone in a land of a yellow- hot sand, another man whose importance could not be doubted just entered a supremely important building on Manhatter, Nork, and his mobile phone rang.

The Nork Museum of Natural History, located on the 79th street (because Nork had many streets), comprised twenty five interconnected buildings housing forty six permanent exhibition halls, research laboratories, and a mammoth library.

"So why don`t we just say that it`s such?!" Shouted the very important man into the

piece of plastic he was firmly clutching in his right hand.

"It`s not that Such, Jeremy." Answered the voice on the other side. "Do you mean to tell me that it`s different? I`ll tell you what it

2h/Footnotes - Continuation several pages After. Sexist language above.

A short userpt on Trob

I looked down to where she was lying, my Magpie. She looked horrific when I found her by the roadside; injured by the incessant hum of cars, going and coming out of the town. It might well have been my neighbour John, going on a scrimmage early in the morning, and hitting the bird as it was packing on my magnolia outside my precinct. I picked it up and cuddled it as it was wrapped up in a large woollen sock. 'If I were You, I would flick it straight to McCullions!' Remarked Sergeant Murphy, with his grin. He was impatiently towering next to me, and looking intently at the bird. Apparently unaware that Lucy, as I nicknamed her, was already getting better.

Sergeant Murphy was my regular visitor. He would come unlooked for and unwanted every time I least needed his assistance. He seemed impatient today, but strangely in a good mood. He kicked a couple of times in the wicker basket where the Magpie rested from my grip and shook his head. People in this country don't say neither hello, nor good-by. He opened

his sedan, four-wheel, and got in. He had been gone before you could say
She sells she shells.

People liked me in this region. And my neighbours always helped me with
so much. McAlice came the other day in her range-rover and brought me a
jumper. She didn't know that I had already purchased four thick blankets
downtown to brave the winter.

125 pages the whole text..

---Page 259------------

On the Count of

Three

I came from The Czech Republic, an utterly strange country to this one. We hardly got out of socialism since 1983. The Irish had a habit of not knowing where the Czech republic even was. I soon gave up on a lot. This Fianna Faill nation An Pártí Poblachtánach seemed to be ruled by people who possessed a wize judgement of things, year by year the land blossomed with beauty of such magnitude that, just to live and work was a gift of sheer blessings, and to have a garden with orchids was a thing of another blessing, the way we had who had their grandparents buried in Minnesota, or Oklahoma. The whole thesis was a crux to a comfortable spree-life. Though here in Dingle \do, the summer resort for Jarmens coming incessantly downtown for icecream, it seemed little bit too biased. The rules are strict here Helen, you gotta fucking mind your job, Antony said to me once. He was my closest neighbour, just across the road. A mason by trade, and an Italian expatriot, he did wonderful teracotas with Celtic motives. I despised his craft so much that we became good friends. I wonder what it is, this mutual hatred to your fellow man Lucy, I asked once half-myself, standing on my 2,45 mile stretch of beach with a boat perched in the middle. We seem to be understood only in locomotion.

I worked for a half-year in Cork. A city so opulent that when I first visited it, my eyes got this strange wet feeling and my throat remained speechless. I worked as an accountant for the local firm Hardis & Son. Occasionally earning my extras by basking with a couple of musical instruments in front of pubs and brothels. Nothing mattered here. All the history of our nation, with its writers, painters, art critics and architects, seemed strangely obliterated by a bubble of idiocy and timelessness. I ate at dinnies, a place for the derelict and despicable.

I' here say that some interesting aspects of Ire we completely put out of my mind as I scummaged for a buiscuit to understand that I could eat. The piece of wet dry food with a sweet taste was enough to saturate my hunger into a disillusioned oblivion. Where is the beautiful day, that I saw in spleandor rising, and I understood that we live for the education of children. I thought that my heart was full of beauty and submission towards natural beauty. To say that natural misgivings are aspects of nature that are visible, you would have to understand that that is only trash. I put out the bin at 7 in the morning (sa wagin {phonetically transcribed}- gealige), I circled my gaze towards the road and understood that it was calm enough in the early dark to pertain such a beautiful day with mundane task, like cleaning and sweeping. [(1. Šuká) Czech word meaning sweeping in old terminology. Modern meaning [fornication / problems with unbiased aligations against certain {with less or hightened Continuation several pages down..

Film competition I.

The film is based on an idea of education, elation and jazz or R&B production. I'm singing west Virginia

pride in a paradigm - those are mormons who travel to Africa to educate and spread belief.

I'm coprolitic in my expression / Book 1. as well, - and I claim that I'm trying to use swearwords and demuring all-recital language- to solve racial problems.

. .

125 pages the whole text..

--Page 260------------

On the Count of

Three

Famil
.
 Enfant

. .

· The film is devided into ½, 2/3 segments (famil, enfant / family, children..

Melodramatic, Allen Petterson opens up

The film is based on minimalism, almost Dada, + new aspects for education / science

Structure of the Narrative

Me - He

You- She your partner / the amalgamation, stilization of a Narrative language is a matter of polemic.

In true deep therapy. Ends up in Aristotel's cave. Normal Oidipus.

Two gods, two religions, two directions of submittion.

After Shock feelings (You constantly search for an excuse, omission, or an excuse that it's going to be better. That feelings of danger will be left behind. People are willing to resort to incredible ends to reach peace.

You can cherish bad feelings on the bases of your language inedequacy. Things feelings should be named,

pinpointed so that they are understood.

I tried to borrow scissors - I was several times told that it's impossible. I appologised.

Pertaning mindwork

· that you eschew from eating something and you remember the past experiences. -

During deep therapy and before after, - you should be warned that you should eat such that yu are only

happy -at the same thime ypur behaviour has to be, or should correspond with mundane.

· I here think .- I'm interested in behavioral arythms / enclusively and collectively called rythms -

possitive (inclusive exclusive negative).

I undergone a fast and in chill I recorded a song that was supposed to be in geanra #rock more affinite to shanson, classical in Aretha Franklin

I hereby disclammer all possible notions and evocations that I use any inapropriate versing (and hereby would call any such presumptions inapropriate, absurd and prefabricated. I'm solving if surroundings and instruments do tend to teach you almost manners.

I almost claim that harmonies -. and proper harmony stearing can lead to a betterment of verbal or

{speech functions. That's why we understand why education is so important.

I would be almost willing to theorise that a musician whu atteins a very good learning of ragas; {upanishadas, and mastery panatipata suta..

can better his/her speech (automaticaly, in succession), we is three, food. 1.Study of how to attein an

instrument

125 pages the whole text..

--Page 261------------

On the Count of

Three

Theory.?/ People nowadays explicate on behavioral a-rythms with the possibility of reaching spiritual climax.

I claim these aspects impossible and call them illusional [disilusion, on e of the roots of evil / Buddhism. ----------------------------------

I tried several times to express profanities, that lead me to understanding that I was in my fast cold and shaking. People with belinguality have much greater perceptive and receptive knowledge

Behavioral extensive rythms [positive, normal, negative and based on general good education, religious codexes of speech, behaviour and job attandance. Person should be understood that he / she can live in a mutual respect /according to religious codexes, good manners, education, and the prospect of atteining a payed job, with + atteining at least one instrument.

Problems arising from inundation

Thesys underline, logical prognoses in afternotes:

the problems ensuing from this as possible behavioral changess, understanding of certain patterns of speech and vocabulary insuing in a pattern malady / pinpointed logical aspects, bases for educational, religion codexes behaviour, upbringing and growth of a man (individual).

I here claim that suck a cock an get a horse! (breton language) That my mind should blow! Is a logical faximily of a prelude, a hyposentence for a tonguetwister, new conage of a fraze - Ensuing further reel on music and education. I claim that coprolaly (unhindred utterance of coprolaly, can better your speech patterns and make order in your vocabulary, fluent your speech and logical aspects of it. --------------------

This suden ubruption of a mood I call Benjamin, Parents ever growing interest in the life of their adult children. Benjamin Suck a cock and get a Horse! Do you like it, I dont like yhe way you hessed up the toaster. Interupted sentences, no links between a fluent logical flow of a language as crux, and the flow of a conversation + normal obsedance of an exhausted conversation with happines flow. ------------------

This simily of a cock sentence (a trigger), a parallel to a family life/rebuff on a logical sentence of your parent. Abundance of possitive thinking, destruction of dreams, and links on a logical feel, and rhythm, understanding of a day. Exasperation, ensuing frustration and envy. - Further/ logical peacing together of logical aspects connected to moods, images, and feelings.

Doesn#t have a day. A whole day feeling through, logical prognoses on a walk out, visit, laughter and normal feeling and channeling happiness, without hindrance or logical prognoses.

Famil / Contrary to that - Life in a Family of parents. Felling safety,
peacing together of logical prognoses,

its often shatter, wave line _/-_-_-_-_- , _____-___

----

125 pages the whole text..

---Page 262------------

On the Count of

Three

I claim that the frustration of current young people 30,32 years old for independent family life is a relatively palpable and strong aspect of the current society, and sociological circles. Stance 1. Appandages:

Filmography is a powerful tool. We have to understand that Caroux and Donstena my plays. Troubles, empty shipyards, 2. Musicology. Jane Awsten. Jane Godall, Jane Srtre..

Problematics of intertwining of mutual prefabricates in mutual correlation..

There is an interesting barier to solve that we would entwine spiritual instruments in music such, that the ether would be visible and would widen to a visible and palpable scope. A similar problematic might arisen in a credit card being million, or billion by minus. There is a strong simily to the current trend of feshion and the possible extrabold on logical live norms. We've always aspired to the ideal of religiousness and happiness, and this notion should always be for mostly exaggerated and put forward for the practice of populi. The problems with everreach into different societies is an automatic flow. We should always behave in the best respect and understanding without harme of the others, in education, understanding and the possibility of a happy life. You should never result to any ignominy for any disillusioned reason as such act is in a contradictory understanding of your framework and structure.

A logical life, religion of love and mutual respect with procrastiny in a relative or absolut happiness without harm.

Jeremy Kile's Show: An interesting show in Ire that represent the current understanding and education of a slightly problematic pairs and individuals, their relationships and their mutual intertwinement [correlation], +procrastiny.

importance]. The person, writer is trying to make known that by the rising of this expression the possibility of such an often unbiased alig. Is raised. We understand that this is a tabu that should almost be abolished from your wrting. Kavannah, slight hints, most preferably about rustic life.

I here understand that srani, at the beginning of text. Is [Czech word meaning feceas] has several meanings in Czech, predominantly stress and bad digestion from stress rising. We extrapolate that bad digestion can lead to the pronoucment /utterance of unheal

Pupei

Puppy, Pie, Pee, Pu, Pay, Pe, Mir Loutka {low, tetiva, ka, you need to go to the toilet when you see a carver. Industry, Uistacht, Usine. Vin Vine, Lumir in Mucha, Lublan. Medow, Poplar + stag,cow, herds/ craft & agricul.

125 pages the whole text..

--Page 263------------

On the Count of

Three

I can explicate almost any word in several langues in several langues by heart.

1. /do . A word meaning incessant empathy, love and possitive thinking without harm. Act 5, everything that is pronounced, more unpronounced and means an utter bliss, without the harm of others or any living thing.

My memories begin by our travel to Italy. I was so small, or wee, as the Irishmen say. That in my gaze

into the past I remember only smetterings of vivid snapshots of consciousness, almost like a coloured reel of a kid that laughs with the happiness of an incarnated glee.

We arrived to a large precinct of our friend, who emigrated to Italy due to previous bad conditions for living in our country. Only now it strucks me how it was possible that my father wasn`t even scared to come to visit him. Karasek. This man, a priest and a preacher, whom now I occassionally listen to on a radio with his songs that pertain laughter among the older ones and an admiration among the young.

1d/Footnotes - Possible parallel and a simily to Domini Derry, my friend. There is a powerful link to the Prague, Czech Pen Club, Arnost Lustig, Viveg, Where the Dog is Burried. The below text almost like The Eternal Lightness of Being.

It was a night like a burning charcoal, and a day like a burned twig, as we ate on a fire and slept in tents. A night by a night and a day on a spar without a deep knowledge of a learned person, I scrimmaged in a bag to pull out a cube of butter, got pested from my mother for the disknowledge of how to spread a peace of butter on a bun. Got photographed on a motorbike of a red colour. An old scooter, in front of this large house like tender. The opulent space coned my senses in my youngster hood.

I never remenber the way back, the road. Nobody understands why we call it a way. That it shows the direction of our travel, and yet it can be a rue. Like a road, a large set of tarmack of a wallop on a hitching car. I remember my mother laughing at the carrricatures of Czechs travelling to Croatia with large cans on the roof, because we were used to purchase a large amount of nutrients to stay in a campsite. An interesting squig of nature, how we had the beach of the south down to split for ourselves, and we got so pested for it. No one told us then, that we live in later, because we lived for it. A rhapsody of prosperity and enough money to buy almost anything we desired.

We colerate, we never do. Smack Ma. Ouch. Ourssss. Usssssss. How are cha. I have never seennnnnn! Snakes on a walloping ice cream, those squig guysssss, sharp steel. Industry. Bookiiiissss. Like earthy colours those paintings. Like the chandelier on a hithing ceiling our our Home. Geal Solas sa tech moi, mo peti.

125 pages the whole text..

On the Count of

Three

I introduce myself, I colerate. I proffer a bussines Card, This vehement building Thached, Tache Tache.. Lare scale architect Build full of crispies. Six houndred kinds, like kind words we exchange, as we proliferate throu the mall. Take it! My dody shouts, she six, or three, exasperated over the surplus.. Take it, fuck! We take several! Large boxess of crispiess. For heaven of several days. We take onions, incoguous things. Buy lunch meat, and hamm, I remember my origin as I take Layagne. I drive with laziness, tiredness, like bitterness. But we call it happy timess, like Kiliwakee burgers. We exchange, I proffeer, swet all aroundeer. We laugh with everyone over the Batch of industry. She tells my savalshe, I immediately love her, and feel indeerment. We colaborate, she explicates as she extricates my shopping. I see shoplifters. They struggle for democracy. I' see them proud as they hitch the shopping, they corellate in a different fashion. My dody thels me I love cha!

Have you seen the Mark O'Shea yesternach, I saketh my Do, as we scurry, on the tarmack, clean polish of a rue, large speed, I talk to my daoughter better then to my wife. We correlate. I wet my eays on the site of the sidewalkii, the tarmack here was born prestine, I greet everyone as I pass with the heavy mesh car, steelen steed, my dody cries fuck you with exasperation and happiness, she never new education, she exhilarates as she knew she study later with me, over a textbook. Mathematics, French, English I teach her improptu, Geography, etc..

2. Footnotes / I aspire to the idea that Antropology, modern anthropology as a branch of science can be explained in a textual form. Ulysess, by James Joyce is a wonderful example of an anthropological, sociological understanding of a nation, nomenclature.

I deem this above text is a good stylystic of a Dada textual form, verging on a free verse. I'm not sure if the possibility of the use, and utilisation of swearwords is even possible in current new literature. I still claim that due to the change in current critique, for a better industry and the look back on the, let's say the literatre of 80s, 90s of the 20 th century, Czech Repub. Might lead to a better view, and understanding of the oncoming ways of writing and modern literature (Postmodern literature in my view. Basically a new beginning like Art Noveau, Art Deco 2035-45.)

Arhhhhhhhhhhhhhhhhhhhhhhhhhhh. Horsssssssssssssssss. Look at me now.. Juju, Pupi, I have 'cha, smeck ma, I have cha, got cha on cha, litricha, kari kari. Toor, like rue.. France Uistacht... Gealge na Loabhar sa rue..

3.Footnotes / The possible extrication of onomatopeia, exchage of word order. Can even lead to new stilistics. (Gealige, Breton, English, French...)

Exchange of word order. Border (bordering on something new) like Labrodie, diew..

.... Look, lok.. I have cha, look at the tv. Po, I've extricated this fayon from under my chest od clothes. Have cha like a palm pad..

Box of match, La brike… Stupido.. Limet…. Like Sherbet lemon..

Lemon grass, Rai grass like Iane Andersoneeen…. Whooo, ye Darrrrh. Ana cha, anoorie, Toyo……. Like Tat… Urpiiidiii… I have never seen some apples being called sherbet. Pommess, [pomes] pomes du terra, almost like

125 pages the whole text..

--Page 265------------

On the Count of

Three

4. Footnotes/ We deem this last textual form an unique ensight into the forming of words, it's changed structure and altered meaning.

I remember my second visit to Italy was a cognomina's act of a surreptitious escape into a land of freedom and living hood. I understood a lot of my Erasm in this respect.

Like mausam he, that mind changes with a scope of recognition of a foreign populi, I understood

immediately my goals and my logical inclinations that were prestine. This holiness of the structured self was widely mirrored by the local interaction.

5. Footnotes/ The eirport to Bergamo, just as the Roma Fumicino was like two roads into consciousness of what Italy really represents.

Rome is one of the most beautiful places I've ever visited. From the Airport I took a taxi to take me about 20km to Rome. The hotels are beautiful and full of good people. I felt the vibe of Milan touching me and the memories of Dominika Dery was so close, that I was keen to see her soon after I would return from the Monastery of Santacittarara.

Next day, I found myself on a Bus stop, slightly confused about my future wherebauts. I have never lost enough hope not to see a good man showing me the way, or instructing me about how to get to a place in an absolut comfort.

The day is like a La, like is the day that takes me through this avocado of haystacs and pastures just as the bus hit's the road beyond the scope of the city. Stelions and wild horses juggle on a dearing in a wild country so prestine, that the mood rises to a prosperous voyage to see old friends.

I arrive and shake hands. The Santacittarara Monastery is woaven in mystery. Ajhan Otala, (Ashin) is a

senior monk and the highest representative of this place. We talk a lot about everything (2).

(2) The short history of almost Everything, just as the Hitchhikers Guide come to mind.

I'm willing to meditate. Love 'cha, we mediate. A medal on cha, love cha.

I depart to my Kuti. A meditation box, cubicle, a relatively large and heafty abode, a structure in a close vicinity of the main Architravee. I sweep, greet with people. Understand. I'm on bussiness. A bussinessman. I have never exchanged love for something else. I have always been enough dearing to have a lie to give the people enough scope to digest what I am.

6.Footnotes/ Digest what I am. An Antropological thesis.. A Frase never seen. A possible existence in future books, and literature of fiction and interest geanras. A joke. (Metafiction).

I mediate sa vagin teach, to crawl to the main house for a grab.

125 pages the whole text..

---Page 266------------

On the Count of

Three

7. Footnotes/Not even slang. Expressions mediate to corelate with reality and the illusionary narration.

I illustrate the Papa monk his Vagin. The morn is good enough like mici fuc-. We love. I love cha sais the monkes. Have cha sais the he, and we mediate.

I proliferate to the othe room on France Tab. We eat cha, we have so lot of it, that harasse over the gram like wee. I have never seen so much grab in my life.

I mediate to the off build where my obode are. I understand the day is about rest. I swovel in my bed. At darkness I crawl to the dining room to aske for foodi. I'm scared to be even given anything and receive lots of giftooirs. [have to read foneticaly].

I understand the late night meditation is on a verge of happening. I go to meditate. Full of ripe judgement I meditate like Moor, or Foramen. They reel and shoot as I meditate. Ajhan does not have so much rep as I swerve the nights. Ajhan Supito I give occasional murky glances. He is just behind me with his rep. But he lets me be reeld. I easyfy and ajhan preluds some monochromatic like a swivel machine. We rep a lot like rascals.

8. Footnotes/ rascals. A possibility of omitin the word would lead to disinterest and not understanding the sent. In a laughter. Like incandescent children.

Foxy.. I have cha.. Tri. Bit by bit, I live with cha. Quadrupa.. Fourscouriocha.

9.Footnotes/ over passing in logical judgementation. buddy is… It's fuuuor

Such and that's the end of it!"

Jeremy Walrus lifted his index finger on his left hand and, pressing a button on his mobile phone, he interrupted the call. Walrus was a handsome man. He had short, raven hair and not being exactly tall, he wasn't small either. Neither was he chubby or boring. That's what he was like.

Jeremy Walrus looked around the immense entrance hall. It was early in the morning, but the sun was already filtering through the high windows in sharp beams of orange light, dazzling the eyes of a scatter of early visitors, stumbling like casipees into the cool hallways of the museum.

"Domestic problems?" Said a voice, and Jeremy Walrus turned around. The voice

belonged to Austen Parrot, the head of the department of Paleontology.

"Parrot!" Chirped Walrus. "I have to embrace you!"

"Look, Jeremy. If you want to start persuading me about what I`ve

just overheard?"

"But it`s Such and nothing in between!" Interrupted Walrus.

Parrot took off his magnanimous spectacles and measuring Walrus with his
half-blind eyes till Walrus looked more like a freshly picked radish, he
finally spoke. "Jeremy, if you need to take a couple of days off. You,
see. Just to get out of here. I`ve always thought that you are a bit like
me, rather seeing a stuffed animal, than the living one you have out
there."

125 pages the whole text..

---Page 267------------

On the Count of

Three

Walrus stared.

"Anyway, I`ve seen you in the lecture room! I think that you are alright there." He

finished his sentence and gave Walrus a friendly wink.

"Mister Austen." Interrupted Walrus still radish, "But I was talking to Seaman. The MD7 have found something. They don`t know what it is, yet. It`s submerged, at least, ten meters deep under a thick layer of granite. The earth x-ray showed just a blur. I think it might be a Brontops, or even a partial Apatosaurus. They can`t judge the age of the layer, yet. Look, Parrot! It's Such!"

Parrot made a long sweeping stare around the hallway and then looked back at Walrus, as if searching for an imaginary prop. He was getting old and senile. It had been a long time since he started losing track with the latest scientific methods and goings, spending ever more time in his study playing with plastic toys of dinosaurs or delivering lectures to women

1b/Footnotes /It's interesting if you as a critique would take this sentencuation immediately as dearing. You have to understand that Palaeontology, Botany, Anthropology / the branches of science generally are only just for a few decades being religiously extrapolated. That we only in recent books find these parenthesis on religion, exploitative on religious understanding of scientific possibility and religious demurement.

interested in the rudiments of Paleontology.

"Look Walrus," he said, searching for an excuse.

1C/ I spent half a year on an Ire Beach, Quartenon 2547. Surrounded by seals, giant

walruses, sharks and dolphins + whales.

Wesk Cork, Co. Kerry, Dingle - Allan P. Natural Reserve. Dingle, Kerry. Is claimed to be one of the most preserved piece of land in Ire.

The forces there and the local habitat corresponds almost to high tides, over heatching

of small species, and the possibility of high winds, warm wint..

"I have a lecture at ten. Speak to that new anthropologist from Mexico. That Gypo

1a/Footnotes - gibson lespoul, gypsy, gypsy King [musician], giro, 2, ro, Roma, east, Po, PE, PIE, Pey /Bread, Peace, dark, cake, spin..

There is a possibility of this word altering its meaning for a completely different thing in the 21 st cen.

can barely speak English but proves to be quite a miracle. Have a good
day."

2/ Here I ostentatively give a notion that I might be from a gipsy
family. Originaly possibly from south America, India. Bhenimin Allan
McGobhain. The one who holds a rock of prestine beuty and can be its
maker. Tolkien /

125 pages the whole text..

--Page 268------------

On the Count of

Three

Aldus Petticoat was an old man. He held a Ph.D. in Astrophysics, Europeen

3a/Footnotes - Europeen Almost like the Ulysess by James Joyce [Ireland Gealtacht, Lunecy.. Possible overlived express. From FF,Uk. Newver heared of in Ireland in a daly conversation/

Linguistics and a completely new science dealing with predicting your own death by

the luminosity of morning suns.

3b/Footnotes - The winter and summer equinox. Celtic hollidays, and celebrations. Mount. Tara, Giant Causeway, Chusulain May, Sleav League.. All visited by me. Benjamin Schmidt [Belingual \children education, growth and mutual corespondance in groups and the society, company, living hood, work a Job maintanace and monay intake, prosperous life in happiness and harmony according to the highest trends.

Trends - Low[e] intrication, entwining into the group, family and the society by undermining influences. From within by missunderstanding of logical trends, from without by forcing an individual into a different scope of life than which corresponds with his/her its individual talents, logical scope and understanding [inprint] /Melevolent influences, ensuing disagreeability, unhappiness.

Normal A relatively standard way of life + with a relatively logical presumtions,

understand ability, possibilities, ups, and downs, and logical feeling boundaries. [I'm a very strong seeker of a Buddhist notion. There is a lot of very strongly religious people in Lunetic Asylums and among the impoverished children. I claim that the bases of this notion and its groundwork can only just have one offshoot for a possible disease. Malady [minor problems usualy in cognisance and plus with born in disabilities. The more powerful is a problem of the inability of the social circles and the surrounding peer group to give an individual enough knowledge for its religious ardency, its vents, logical extrapolaty and understanding + respect [for an individual].

Highest a possible large money intake people, nonsensical performances on the person from the side of the peer group. Threatening, blackmailing. The rest like Normal.

Thus, while the completely unimportant, hungry, truck driver Jordan Bruno was wiping his eyes into a smudged overalls and while a very important Jeremy Walrus still radish stared at Austen Parrot who measured back his outcomes in lectures in a slowly-filling museum entrance hall, Aldus Petticoat crawled out of his bed and looked out of the window.

"Look!" He said to himself while staring at the gigantic ball of light.

"I`m gonna die, soon!" He mumbled and averted his eyes.

Aldus Petticoat walked into his study. It was a room with walls made solely of

125 pages the whole text..

--Page 269------------

On the Count of

Three

bookshelves. He wasn`t searching for a book, though. His hands were trembling and he needed his breakfast pill.

Petticoat was a Wer-renown capacity on many subjects, but he was also a communist, hypochondriac, and a complete fool. His foolery being probably best reflected by the pill he had been searching for on this peculiar morning.

"Where`s my wallopy` medication!" Cursed Petticoat while frantically rummaging

through the drawers of his writing table.

The pill had been prescribed by Dr. Gerard Death, one of a few people whom Petticoat still trusted. And although there was little reason in trusting a person with such a name, Petticoat nevertheless took his pill of pure glucose diligently every morning to drink down with a cup of mint tea prepared by his maidservant.

"What are you looking for? Asked Mary Slaving, entering his study. "My

wallopy` Death pill! Answered irritated Petticoat.

Petticoat never married and Mary Slaving had been the only woman who had ever crossed the threshold of his capacious household. They had many things in common, one of them being their unchoosy tongue for food, and the second their unscrupulous behaviour towards the outside.

"It`s right on your table!" Cried Mary Slaving, seeing the big, white pill

glittering in the sunlight.

"Where?!" Hollered a frustrated Petticoat.

"Look, there it is!" Shouted Mary Slaving, crossing the room and picking up

the pill.

Petticoat put on his spectacles and observed the pill - a big, white, glittering object in Mary Slaving`s hand. Petticoat did not believe in any religion, but taking the morning pill was a rite he could not go without.

"Now, bring me a nice` cup of tea, or I`ll kook something fer cha-!" "So, I'll make you some pasta-!"

1d/Footnotes- I here in the last couple of sentences exchange rude and profane expressions [that might sound hilarious, Charles Bukowsky, Pulp, Even Joseph Heller [The Porn, or Trainspotting by Irvin Welsh is a completely different geanra for canversation or literally endeavours], even a word Nice can sound rude or obscene in repetition or different wordings, phrasings or locutions].

1e /Footnotes - Fi Fe Food - is very Fren.. Mauze, like Mausam he, Hindi.. Jidlo her feed dloub, wallop

or a spoon. A man feeds a woman or a child.

Fo, fo, foood…

It was ten in the morning and Aldus Petticoat was sitting at his book-overfilled desk. It might well be said that he looked serene, but inside of his skull thoughts of incoming danger were ever present.

11/a Footnotes- Aldus Petticoat suffered from a chaseing

desease, or persecution as an important man. He was a

125 pages the whole text..

---Page 270------------

On the Count of

Three

disillusioned individual maladed by years of teaching and study. I would say that a similar case is possible to be found in real life, a problem of overeducation without enough justification. I call in Charles desease. Without any logical background of affinity.

For the last ten years he had been trying to put down his memoirs, inclosing himself in his study for longer and longer and even longer periods of time to ruminate over his life and, of course, what comes after it.

12/a Footnotes- A tipical disorder now often found among middle age people and singles, who feel they have evergroun the possibility of having childen Husband/or a Wife, an independent family life. Truly existing and arising in people who are just in a phase of would like to obtain independence in a new relationship. They often break down befpre finding peace in their loneliness, or reclusiveness [seclusion /Film 2, Couroux].

Petticoat`s worry about his life grew day by day, and so grew his worry about ever finishing his story in time. His memoirs started like a novel. Then, about five years ago, he decided a short story would be preferable. And now, the last five pages were written in shorthand.

13a/Footnotes- A tipical problem with concentration of an enough duration not to alter your plans for the longness of an essay, holiday, a normal day activity, problem with laptop ediction [fluctuation syndrome, computer take in].

Fluctuation /walking aimlessly for a logical goal knowing that its just a lure or a stratagem to maintain, retain a logical sanity of the day, subconsciously knowing its deeper purposlesness, ensuing logical superfluity of days with a very high happiness stigma feeling. I call this syndrom Evelin. I do not think that the treatment of this desease or syndrome should reach beyond the scope of accepting parents and educating the whole family of the importance of the individual suffering from this diorder.

125 pages the whole text..

---Page 271------------

On the Count of

Three

His/her family should be informed and educated about the needs, scopes and views of this individual without the individual's interaction with them. Putting the individual into the learned family which should not further ostracise him/her with nonsensical questions about education, spiritual codexes, or belief. Such individuals suffering from this disorder should be received gifts, a notion that they might change their names, get christened without their parents presence etc. The religiousness, and spiritual codexes of such individuals should never be doubted, they should be asked questions hinted towards their religions, education, and should be admired for the knowlidgebility of their answers. They should be almost helped in obtaining a partner rather then just being instructed how. The syndrom is most often palpable among adults of 28-32 years old but can arise in even younger people or in late age, anytime a person looses his/her partner bond.

"Duck." He whispered over a blotch of ink.

"Nah! I can`t carry on like this. I saw the sun. I can for sur depart tomorrow!" He

leaned back, and thought…

"Perhaps, I should make it a poem."

"Since I was a little brat,

I slugged hard as a gray a rat [topo]…" "Yes, yes…" 5/uFootnotes- Slug - similar to snail, Couroux Film festival

--

The Nork Museum of Natural History had a special building in the vast courtyard of its complex. It was a ten story building known as "The Ritz". The Ritz did not receive

its name for nothing. Ritz was notorious for its large and well equipped research laboratories, lecture rooms, and also very clean toilets. I mean to say, it all used to be. One day, the main part of the museum got simply overfilled with stuff. The corridors were lined with boxes of bones, back-bones, and bones that did not even look like bones. Until the situation got desperate. Trucks of

125 pages the whole text..

--Page 272------------

On the Count of

Three

unwanted bones were parked outside the museum. There was a crate of bones on each employee`s desk, and unflushed bones in the toilet. Employees themselves did not go for a cigarette

6/Footnotes - Cig /a/ret gipsy, black and white, cini hindi/yellow {and white] Couroux film comp.

anymore. "Hey, Charlie, let`s go for a

bone."

14a/Footnotes- Antropology is a science very important for a parallel growth of sociology and has a strong impact on a lot of branches of science. By the study of bones, we can lear a lot about maladies and thereout a lot about the past life, manners, ways of living, we can simulate and mimick old life patterns for a historical efection.

And they would go outside, each of them chewing at least two bones before they had to return back.

But there was a man who saved the day. He was a small Hindu, called Dada Chapati. He worked in the basement which was the most afflicted area. And while he was having a bone with a friend of his, Chapati got an ingenious idea. He dressed himself as a carrier, and humping a heavy box of bones, he rang the bell of the Ritz during a lunch break.

"I have an agent post to be stored in this building." He announced in his broken Englis

7/Footnotes- broken language, belingual children with an abnormal happiness feelings steming from high education are in fact very healthy children, who can ask for work prosperous life and living hood.

to a confused scientist who answered the

125 pages the whole text..

---Page 273------------

329

On the Count of

Three

heavy, glassed door. "Well, do you have a conveyance bill? Asked him the
scientist. Chapati was ready and he proffered a piece of paper carrying
traces of his breakfast. The bill had been written in Hindee and said:
"Hereby you say that all the bone equipment of the museum can be stored
in this building. And nice you! Your mama has so nice feet that she has
to tear trousers!"

14/b Footnotes- You would be surprised that this insult can in some
countries be one of the worst insults you can ever say, and therefore you
should never use it. It is logical but not often understood that we can
buy a swearword, slang dictionary to see and understand certain
sentesuation, wordings or slang,. You should be nevertheless aware that
the use of profane language in publick and most geanras is almost
forbidden due to a common sense and international rules of writing.

"Furlow!" Said the scientist. "I can`t read none of this, but it must be
coming from very far!" Then, he signed.

Chapati returned to the main part of the museum whistling. Since then,
the Nork Museum of Natural History became the only public institution
where you could actually whistle.

15/Footnotes- From the past History in many countries around the world
we've gleaned the information that some people are willing to ostrecise
others for being more happy than some other people.

For example in the Cyech rep. it was

called Iron curtain.

Chapati was given a rise and a new position. He became the sole scribbler
of conveyance bills, while the previously indomitable building in the
courtyard was getting overfilled. The museum began filling

125 pages the whole text..

---Page 274------------

On the Count of

Three

out, and the inevitable doom of the Ritz was thus sealed.

Jeremy Walrus entered the Ritz and walked into the elevator. With all the boxes of bones around him, there was just about enough room for him and perhaps, well, another box.

As he pressed the button to convey him to the ninth floor with the index finger of his left hand he had one single thing on mind. He had to speak to Gerard Baboon, a leading expert on anthropology in the whole Wer and one of few people involved in the project in Sierra Tralala.

The last two floors of the Ritz were the only survivors from the golden days, having still functioning laboratories occupied by a scatter of scatter-brained scientists.

The elevator door opened and Jeremy Walrus stepped out into a long corridor. Gerard Baboon stood in front of a coffee machine. Since early in the morning he had one single thing on mind. He desperately needed a cup of cappuccino, and he had to have it presto.

"Blasted engine!" He cried in his bad Engli. Broken like the bones all around him. Jeremy Walrus approached.

"What`s up Baboon?" He asked.

"That boldy engine dust not fork!" He answered to Walrus and started fingering the slit for the return money.

"I have to speak to you about MD7,

Baboon." Chirped Walrus.

"Give up, you Betsy!" Mumbled Baboon under his breath, his fist tired. "Do you have any news off MD`s?" Asked Walrus in a tone of business.

125 pages the whole text..

--Page 275------------

On the Count of

Three

"Yeah, I have them on the fire, right now. I just felt like a breach, wanted my cappuccino, Wallthrush." "Walrus" corrected Walrus.

„I appopledgesay, mr. Wallrush. Anyway, would you

like to falk on the fire with our theme?

"That's why I've come'ere, Queens." Said Walrus.

It was fifteen past ten and Austen Parrot walked into a room full of people. Whether he knew if he walked into a lecture room on the fourth floor of the Ritz, or into a pavilion full of breathing and heaving bipedal Rinoceroses, no one will ever know. Austen parrot lived in a completely different Wer of his own. Still, he retained the dignity of a scientist and a frown of a scholar, which, can be said, never did leave a young lady cool.

"Welcome!" He anounced, and knit his brow to the accompaniment of several happy smiles. "We`ve assembled today in this luxurious room to talk about what life was like when the life that we know and trust was not yet, but to come." He finished his sentence and looked around the room to make sure he was not understood. Then took a deep breath and carried on. "When the Wer was still young and fresh, creatures unlike us walked the surface we now walk, killing and eating each other in many an atrocious style!" Austen Parrot raised his head and made sure everyone wasfeeling good. "Thus, if you noticed that I said the word 'quite' with an exaggeration, I meant that this theory is a nonsense, or better still, an utterance of a scientist full of nonsense. But let's put biases aside," he diminished.

Austen Parrot looked around the room for the third time and knitting his brow to a cacophony of happy whistles and aplaus, he solemnly declared. "Today, my dear all, we will talk about a creature very much similar to us. We will talk, and God knows I have to wet my forhead, about a sixty million year old dinosaur called…"

125 pages the whole text..

---Page 276------------

On the Count of

Three

Walrus and Baboon were sitting in the connection centre on the ninth floor of the Ritz. A room full of beeping, blinking and buzzing metal boxes in racks upon racks upon shelves.

Gerard Baboon was sipping his cappuccino while Jeremy Walrus, now expert, was fiddling with the buttons of a radio-signal receiver, trying to gain a contact with Sierra Tralala.

BZzzzzzzzzz, Zzzzzzzzzzz, Whiiiiiiii…

MD7 was a cover name invented by Austen Parrot. It stood for Mad Norks D, because only a mad Nork would go searching for fossils in Sierra Tralala. Seven then stood for the number of people involved in the project. There were two palaeontologists, two people from the Green Peas Taste the Best competition, who won the trip in a

shopping mall, one student of anthropology, one pediatrist, and the

leader of the whole team, Emerson Seaman.

Whooooooo, Whiiiiiiiiii, Plummmmm…

Walrus stroked the metal top of the radio-signal receiver with the index finger of his left hand, and the voice of Emerson Seaman filtered into the connection centre. "Semen leaking, Wzzzzzzzz." Bleated the loudspeakers.

"Good shop Wallthrush." Said Baboon to Walrus and dipped a cookie into a cup

of brown liquid.

"This is Jeremy! How`s things Emerson?!" Asked Walrus, shouting into the

microphone.

"Walrus! Good to hear you, man! We are just breaking the ice!" Answered Emerson Seaman; while distant noises and background voices interupted his audibility "Come on Heller, you hold that drill like a karaoke Pony."

"How long, yet?!" Shouted Walrus with a tinge of excitement, while Baboon

choked on his biscuit.

"As I said," hollered Seaman. "We are a way into the granite layer! Stay on the

wire. It won`t take long!" And then..

"Heller, you astere man!"

125 pages the whole text..

---Page 277------------

On the Count of

Three

VVVVVooooiiiii, plop, Beeeeeeeep…

"Tell us what`s gong on!" Screeched Baboon, grabbing the microphone from

Walrus`s hand.

"Who`s that?!" Emerson Seaman was enquiring.

"Buffoon on the fire, Semen." Cried Baboon, spitting cappuccino, and then..

Hey Heller… Zzzzzzzzzz, Pony… vzzzzz, We are there…

"Look at

That

Man!"

"It`s a

Trodon!"

"Who are "Holy, what`s you?!" "Bruno!" Going on!" "I "I`ve " !"

Knew brought you "It`s

that!" the rivets!" Big!"

"It`s holding something!"

"Mother!"

"It`s holding a book!"

A new day dawned on the city of Nork and a stray beam of sunlight tickled the nose of Aldus Petticoat. He woke up as usual, with a ghastly cry of dread. The door to his bedroom were open, and Mary Slaving entered, swetting over her breakfast.

"You, you, you! If you weren`t shrieking like a madman, I wouldn`t have dropped the Chinese vase I was just polishing!"

"You Mary!" Cried Petticoat, half blind with

sleep-crusted eyes.

"I have to see the sun. I fear the worst!" Barked Petticoat and crawled out of his bed. Today he

could hardly walk. He slouched like a poodle to the window and then

looked out. And because a story must be interconnected to make it more fun to the reader, it might be said that, at that exact moment, just as

the full blast of the morning sun hit Petticoat`s face, and his cry of
utter happiness filled the streets of Manhatter, a man at a newsstand
just bought a

125 pages the whole text..

---Page 278------------

On the Count of

Three

copy of the morning issue of The Dork to Dork newspaper, showing a six
hundred million year old dinosaur clutching a book, and laughed. Then,
very, very far away, a Buddhist monk named Opal just opened his eyes from
a deep, deep, concentration and realizing that what he had been searching
for did not exist, he decided to skip lunch and take a walk. A Hindu,
employed in the basement of the Ritz, and named Dada Chapati, just
invented out of boredom a flying machine proppelled solely by bones, and
Jerremy Walrus pressed the doorknob with an indexfinger of his left hand,
and entered the study of Austen Parrot.

Thus, while the whole world was laughing and rejoicing, Aldus Petticoat
enclosed himself in his room and knowing his time had come at last, he
was illuminated with a one last, desperate idea. He opened a drawer to
his writing desk and pulled out a loaded revolver he kept for shooting

pigeons off the opposite roof. He positioned himself in the middle of

20/Footnotes- Suicide can be a terrible way how to get out of trouble.
Depressions in early age should be treated by a very possitive company
and surroundings. Unhappy replationships, unstable family life and debts
can lead to such depressions.

his study and holding a sheet of paper in front of his head and the
revolver behind his scalp, he

pulled the trigger in his last and desperate attempt to put his mind on
paper…

…Bang!

The news about a new paleontological find spread quickly. The evening
issue of The Dork to Dork brought a printed version of a Judy Tatters
interview with Emerson Seaman, headlined 'Breakthrough!!', with two
exclamation marks.

Mrs. Tatters was a wife of a plastic surgeon and she made her career in
journalism thanks to a controversy over her operation on eyelashes,
having them so long that when she blinked, it looked like a butterfly
taking flight.

"Our readers want to know the details." Said Tatters in the

papers.

"Well, Judy. You see," went the lines of Emerson. "We are digging in the
most atrocious conditions. You know what it feels like to supervise a
bunch of oddstuff and having to wipe forhead with sand."

"Exactly!" said Judy Tatters, fanning Emerson with her

surgery.

125 pages the whole text..

---Page 279------------

On the Count of

Three

"We dig up the Wer first, you see. Then we struck a layer of stone as hard as Eski. We break the stone with drills. There's a sort of chamber and, yeah, that wallop is right there. Not that I have not seen a Trodon before, but they are rare. Anyway, I gotta tell you Judy, when I saw it, I mean when they saw it and then came to tell me. I mean, when I was told and then saw it. I was seeing it and did not believe it."

"What kind of book is it that the fossil's holding?" Persisted

Judy.

"We dunno. The letters are like we know now, if you understand. You see, like those we were taught at school, but the language is forign. We think it might be a dinosaur language."

"Are there any means of decoding the letters?" Asked Judy. "There is a man who is able to decode them. Possibly only he may know, but he is at The Dork Hospital in a state of unconsciousness. I cannot tell you more."

"Thank you Emerson."

Aldus Petticoat was still alive. "We are not sure for certain."

Said Dr. Gerard Death to the curious print.

"The book has been transported from Sierra Tralala and we'll try to bring Mr. Petticoat from unconsciousness." Said Jeremy Walrus, standing close to Death. Twenty minutes later, the complete MD7 assembled around Petticoat's bed. Dr. Gerard Death, although incompetent in the matters of resuscitation, but still the only trusted person of Petticoat's, was also present. Death representing, as it were, the only medical capacity in the room, he rummaged in his pocket and pulled out a pill. "This, my friends," he accosted the crowd, "is a red chewingum. I bought it from a man in the subway. If this won't work, nothing will." Gerard Death squeezed the pill into Aldus Petticoat's Mooth with a frown of concentration.

"1, 2, 3…" Heller, the pediatrist was counting on his wrist

watch.

"10, 11, 12…" The whole room was mumbling like a congregation of priests. "Look!" Cried Petticoat, waking from his limbo.

125 pages the whole text..

---Page 280------------

On the Count of

Three

"Hurry!" Whispered Gerard Death.

Jeremy Walrus proffered the book with a

shaking hand.

"Can you feed it? Blurted Gerard Baboon at the

confused Petticoat.

"Where am I? And who are you?!" Petticoat

bleated at the geathered audience.

"Feed it!" Shouted Baboon, nervous because he

was again thinking about his cappuccino.

"Cook with Mildred in Frenc, you!"

10Footnotes/ the problem of unfinishing words and sentenciation is

similar as in Flour de Mal. The cursed poets are a very interesting and prognostic picture of Pari life where you might suppose that when you travel really far [expatriotic problematic., job attenment, a very good prosperous living hood. There may be a possibility of completely changing the way of life of an individual. Mind broadening and a recursive comeback to basic inner values and understanding. Happiness. A possible attainment of hobboes and children dreams which might have been up to then completely overwhelmed by surface prognoses.

Such child dreams and hobbies are very important to be further developed and helped by others in developing as they represent the most crucial bases and founding stones of a personality and its growth/ A possible problems in modern society may go against these intrinsic values and therefore lead to unhappiness of the patient/individual and its core is represented by the dissilusionment of the sister/Doctor or

practitioner. + possibly trying to find analogy in Parents or even thinking that you are similar to your parent when you are usualy similar to your grandparent of which the Psychologist does not have any information. He/she deems them over excess. Simplification in current science can lead to a mal work. Enough prognostic evidence can lead you to the same conclusions as at the beginning. The current psychiatry and psychology holds the notion and works on the bases of altering you logical rutes and ways, and trying to change the ways and desrupt your logical structures due to an accident. I think there shoud be a mediator, medium way, middle path to go exponetialy from the last two segments askance the way up.

.

. Where psay. Should be at least aimed to..

.

Understanding [comprehension of a ripe personality as even the bases for a session [first intake

125 pages the whole text..

--Page 281------------

On the Count of

Three

of a patient]. I'm an ardent seeker of this notion of practice. Desruption [disrupting basically elements and structures]

7a. Footnotes/ Pictures, happy illustrations and caricatures aim to signify what is hilarious and easyfied. Lofty in a hilarious manner, Laughter-trigging.

II.

Jenifer got up and passed me the salt. It was five in the afternoon and we were eating. I was sitting at the head of the table, my son Tim was on my right and Judy,

125 pages the whole text..

--Page 282------------

On the Count of

Three

my daughter on the left. Jenifer was far, far away. So far, that I hardly saw her features. We had a big table. It was the only big thing we had. We were very poor.

2aFootnotes/Problems with belief, understanding of religion a political structures according to current trends and notions lead us to a notion, Laughter. A possible analogy and simily to a complete corruption of reality is in a man like Kavannah breaking a broom or a rake, shovel and walking away. Then we understand the logical leadout of things, ways. We should still understand that the book On the Count of Three Slam Ben Allen is about Ire. Quartenonb5267 and their religious codexes problematic and, ontake. The problem with learning, education and logical system changes. The result is a try for finding of a new common sense, analysis, and a new viewpoint on the cultural undergrowth. That is, a higher position.

"Eggs again, Mum?" Bleated Tim.

"Be your Mummy`s darling and eat." Said Jenifer.

"Be, bee, beeeeeee!" Judy started crying.

"Very sake!" I said and thumped my fist on the table.

Minor disputes were our daily bread and butter. Literally. We never quarrelled. I was

usually too tired, my wife too good natured, and my children too scared.

6/Footnotes- It's interesting to suppose that the story is solving current problematic of religious biases, and a look upon religion in general. I would claim that there is a possible offshoot of the story in a comparison to Israel and its good living hood. The problems of a family life and the upbringing of children.

"I want ham, Mum." Complained Allah "We don`t have any, darling." My wife said. "We, wee, weeeee!" Sobbed Judy.

"Damn it all!" I said and thumped the table for

the second time.

I was the only working person in the family. I wrote for a small rag called "Shivers". It was a porno magazine showing lots of dirty pictures. My work consisted in editing the speech-bubbles of the protagonists. My parents didn`t know. Once, my father asked me what I did. I said: "If I told you the truth, you would throw me out of the house." He understood.

125 pages the whole text..

---Page 283------------

On the Count of

Three

My wife never worked. She cared after the children. At least, that was what she always defended herself with. As far as I knew, she slept most of the day, or stood in the doorway to our bedroom playing with the telephone wire and talking crap to her half-witted friends.

I met my wife on a business trip to Jerusalem. I was to lecture there on typography. I had

a bloody Ph.D. in that, just so that you know.

I remember the lecture room being half empty with just a scatter of listeners in the front rows. Jenifer was a student then, studying the same thing I had studied. I don`t know if it was the jet-leg, or the pills for high pressure, but I fell for Jenifer the moment I saw her in the audience. She was the only person who remained after I finished. We nurred on the women`s toilet.

Jenifer left her studies and came to live with me in Europe. I was thirty-five and she was just

twenty-one. We spent a couple of years in Venice. That`s were Tim was born.

"Finish your egg, Tim." I said

"I`m not hungry, Daddy." Said he, and leaned back in his chair. "Yes,

you are. You are just being fucking picky!" I said. "Don`t be so fucking rude." Retorted Jenifer.

I strongly disagreed with the name Tim, but Jenifer was Spanish and I thought that she should be occasionally allowed to have her say. Anyway, that idea evaporated from my mind when Tim was just about five years old. I took him to New York for the weekend. It was shortly after the eight thirty buzz of my clock, just so that you know. My son got lost at the airport about fifteen minutes before our flight back was due. I remember running around the refreshment corner shouting his name till the whole place began swarming with cops. They cornered my at a coffee stand and beat me up with plastic chairs. I never felt so much shame in my life.

125 pages the whole text..

---Page 284------------

On the Count of

Three

" Pass me the newspaper, Judy." I said after we finished eating. "Yes, Daddy." She said, and tried to get up from the baby chair. "Can`t you see that she`s too small!" Retorted Jenifer. "Yeah. She`s a wee pee!" Said Tim and started laughing.

Judy was the only person in the family who did not insult me. She was my darling. With just three years on her score, Judy showed all the signs of a future genius. Sometimes I had funny dreams about her; lecturing in an overfilled auditorium on metaphysics, or supervising a paleontological excavation somewhere with lots of sand

around. Sometimes though, when I came home overworked, I dreamt about being still at the office and editing one of those smutty pages. Then he would appear, my Boss. Always on the page twelve, naked but for a leaf. I would read what he was saying and sweat would start running down my forehead. Then I would try to erase those lines, to rub off that dirt out of existence. From such dreams I usually woke up crying and my wife would hold me in her arms, whispering: "It`s not your fault John!" It`s not your bloody fault!"

The problem was that I liked my work, or better still, there was nothing else I could do. And also, I was quite good at what I did. Once I even got a sort of bonus when on a whim I put one bubble into rhyme.

I think it ran like this:

"There is a way like a trad, like road that we walk down."

I remember my boss liked it a lot, and he said something I did not quite understand; "You deserve better than this, John. You really do!" I passed the remark over and worked hard till evening.

Concerning my relationship with Jenifer and the type of work I had been doing, I`m glad to say that I never dreamt about other women. On the contrary. As the years went by, my dreams about Jenifer became increasingly wilder and wilder, and sometimes so far-fetched that I began to fear a brain tumour.

One of my constantly recurring dreams had always taken place in a desert. I was riding a camel with a man whom I considered my friend. Both dressed in black to suit the atrocious conditions we were exposed to; we smuggled carrot soup and peanuts over a guarded sand dune. The desert had been dangerous and we were carrying guns. Sometimes we met with desert pirates and we had no scruples in changing the virgin sands into a place of slaughter. With our camels trotting over corpses and to our gun-shots aimed into the air as a sign of victory, we would then cross the dune to find an oasis. She would be there, my wife Jenifer, sleeping by a water tank. I would

125 pages the whole text..

--Page 285------------

On the Count of

Three

raise her sleeping body in my arms and roar like a lion while my friend
would start franticly shooting around himself, probably stricken with a
thermic fever. Then I usually woke up.

"Anything new?" Asked Jenifer as I was listing through the papers. "Well,
they say Brad Pite got drunk yesterday night at some private venue and
danced naked on the table." I said and tried to focus on the blur of my
wife.

"Anything else?" Said my wife, spooning remnants of a boiled egg into
Judy`s

mouth.

"Yeah! I`m reading here that they are looking for editors to join the
staff at The

Bitter Teet.

"Not in front of the children, John!" My wife exclaimed. "But Jenifer,

It`s my work!" I tried to defend myself. "Teet!" Cried Judy.

Dreaming aside, there were two things I hated in my life. One was when my
wife started swearing in Arabic. The second being private parties held
monthly by my

colleagues. At first I didn`t know which was worse, but then I loved my
wife, so I transferred all my hatred to those sickening evenings of glee
and supreme stupidity. The problem with me was that being in the porno
industry for so long, I had so little self-respect left that I always
went. No, really. It was much worse. The evenings were not mandatory, but
I turned up every month because I was scared. I was really scared that I
might lose my job.

Such a private party always took place in the office where we worked, on
the last working day of each month. The only demand laid on you being to
come dressed up to the nines and bring as much alcohol as you could
carry, or afford. That day anyone hardy worked. The tables were laid up
with booz from early morning and it was hard not to see the gluttonous
co-editors sweating at the sight in front of them. At the end of the
shift the boss usually appeared. He gave a small pep-talk and then shook
hands with everyone as if he even knew our names. The party began.
Everyone was drinking beyond their limits. Men, women, even dogs that
someone always brought to show off their pet`s pedigree. Everyone was
boozing like they were twentyone.

I always kept to a corner. With a glass of red wine in my hand, I felt in
relative safety. Sometimes, the evenings were just about alcohol and
talk, but sometimes, it turned real nasty.

I remember a particular night when one of my colleagues took the pain to
bring over a stereo. I was just standing by a canapé table talking to
Frederic Bruno whom I briefly knew from the office.

"So, you got a Ph.D., eh? He stammered from drunkenness. "Yeah, might be said." I answered half apologetically. "So, ya got, or ya gotn`t." He persisted.

125 pages the whole text..

--Page 286------------

On the Count of

Three

Then it happened. Someone put on a CD with Ray Charles and his Georgia filled up the dense atmosphere. I don`t know whose idea it was, or how could I have possibly been forced into something so base, but at that moment it seemed a good way of escapeing from Frederic Bruno. A human train of my drunken co-editors was just passing by and a fat guy named Bulvar Rolst pulled me in.

We made a couple of loops around the office and then broke through a door leading to the printing-studio. The studio was full of people working nightshifts to prepare the next issue of "Shivers" for a morning release. You should have seen their faces, or perhaps you shouldn`t. Then someone switched off the light. I returned home that night all bruised and with my suit torn at the armpit.

" Go play to your room." I said to my children after dinner. "Don`t tell them what they should, or shouldn`t do." Said Jenifer. "Look, I have to talk to you, Jen." I said, and gave her a wink. "Alright kids! Bed, now! She exclaimed.

When I gave my wife a wink, it always meant the only thing. The very thing. I don`t know who started this game. At first, I thought it was Jenifer. Back in Jerusalem she had been winking at me the whole time. At cafes, restaurants, or even in elevators. We socialised wherever possible. Only in later years she told me that she was allergic on disinfectants.

We stood each at the opposite side of the large table. Finally we were alone. Jenifer said nothing and started cleaning the table, putting the dishes into the kitchen sink. The telephone rang. I went over and picked up the receiver.

"John Stone`s family." I said, introducing myself in my full name.

"What? No. Not Flintstone."

"Nah. I`m definitely not trying to befuddle you!"

"What? Are you offering me an editorship in The Bitter Teet?!" "But,

of course."

"Good bye."

I turned to Jenifer. She was standing at the sink with a plate and a kitchen rag in her hands. She was looking at me intently, as if half reading my thoughts. We both looked at the table.

125 pages the whole text..

--Page 287------------

On the Count of

Three

III.

My name is Bill. I was born in an empty, coffee-stained pack of cigarettes, in a small bistro in the world of Not Much. I would like to relate to you a story of a night so bizarre that whenever I remember it, my backbone experiences a strange tingling sensation not dissimilar to a spinal orgasm.

But first, let me tell you something about my childhood. I grew up with no parents. No Mummy, no Daddy, only two sisters to take care after. They did not speak. They only rustled. Something like when you fold a square of toilet paper. Their names were strange, and as soon as I knew how to read, because their birth certificate was painted in big, red letters on my house, I started to call them Puff & Enjoy. Funny, isn`t it?

I used to like to embrace my sister Puff because she smelled of exotic lands, but I did not like to hug little Enjoy. She was chubby, with black hair, as if she did not belong to the family, and smelled like burned tires.

Me, and my sisters had a nurse who often helped in the kitchen and even cleaned the garden when people left after a short party of high-brow talks and coffee-abuse. We liked her a lot and called her Evelyn.

Evelyn was my teacher. She taught me everything from my first letters and love for music to Kama sutra positions. All till the age of eleven. Evelyn was Spanish and her English was bad. Well, in fact, it

125 pages the whole text..

---Page 288------------

On the Count of

Three

was worse than just bad. She was mixing Spanish

and English, and who knows what other language because her father was a
Navaho Indian… That`s why my English wasn`t worth a slobbering fox… Then
I woke up, of course. But you might be asking an interesting question.
What has changed when you opened your eyes, Bill?

"Not Much"

It was Monday and I was sitting in my couch smoking a lazy cigarette
after a good nap, while my window was passing through a heavy-duty test
of endurance against natural catastrophes. The month of August was in its
most atrocious episode and I wished that my part in it were at least for
that day cancelled. While torrential rains were sweeping down the street
and something like hail, which through all the murk looked like black
coal, was smashing against my single window. The TV set was flickering
and I sincerely wandered if life had any tangible purpose beyond getting
up in the morning, taking a shit and going to sleep again drunk with a
cheap bottle of wine from the grocery on the corner. I got up from my
couch and stabbed the burned out cigarette into an overfilled ashtray. I
had many ashtrays, I thought, while looking on the pandemonium outside.
Out of the whole collection I particularly liked the one I kept next to
my crapper. It was small and shaped like a swan, and according to the
half-insane Chinese I bought it from, it was made out of some special
stone. The China man`s English was even worse than mine and I think he
said something like: "I would not sell you a fuck, mister". So, I
naturally hesitated and then took it home.

Rain was pouring down the Fitzgerald Avenue. It was four in the afternoon
and all the lights in my room were out. I was still standing by the

125 pages the whole text..

--Page 289------------

On the Count of

Three

window when the telephone rang. I went to the table and picked it up.

"

H u l l

o " ? "

B r

o t

h e r , i t

` s

m e .

" " I

d o n ` t

h a v e

125 pages the whole text..

--Page 290------------

On the Count of

Three

a b r

o t

h e r ,

u n l

e s s

m y f

o l

k s f

o r

g o t t

o t

e l l

m e

125 pages the whole text..

--Page 291------------

On the Count of

Three

s

o

m e t

h i

n g .

W h o a r e y o u ,

a

n y

w a y ? " " I t

` s

m e ,

G

125 pages the whole text..

--Page 292------------

On the Count of

Three

a r r y ! " "

A h ,

G a r r y !

H o

w ` s

t

h i

n g s .

I

h a v e

n o t

125 pages the whole text..

---Page 293------------

On the Count of

Three

s e e

n

y o u r

a s s

f

o r

e t

e r

n i t

y .

"

G

a r r y

w a s a g o o

125 pages the whole text..

--Page 294------------

On the Count of

Three

d f

r i

e

n d o f

m i

n e .

B o r

n c r i

p p l

e

d ,

h e

m a d e

h i

s

125 pages the whole text..

--Page 295------------

On the Count of

Three

l i

v i

n g b y r e a d i

n g p o e

m s t

o

w h o r e s i

n e x p e

n s i

v

125 pages the whole text..

--Page 296------------

On the Count of

Three

e b a r s a

n d c a f

e s .

H e

h a d o n e

h a

n d o n l

y ,

b u t

g

125 pages the whole text..

---Page 297------------

On the Count of

Three

i

r l

s s

o

m a

n y t

h a t

o n e s

o o n l

o s t t r

a c k o f t

h e

125 pages the whole text..

---Page 298------------

On the Count of

Three

n a

m e s ,

a

n d

w h e

n h e s

h o

w e

d u p o n a r e a d i

n g h u g

125 pages the whole text..

--Page 299------------

On the Count of

Three

g i

n g o n e ,

y o u d i

d n o t

k n o

w i f i t

w a s

M a r y ,

o r

125 pages the whole text..

---Page 300------------

On the Count of

Three

V i

v i

a

n .

"Bill,

I

have

a job for

you." "Wha t is it, Garr y?"

"Bro ther,

it`s a whop per!

You`l l be acco mpa

nying a

poet just

arriv ed

from L.A.

He is a

secon d

Allen Gins berg. "

125 pages the whole text..

---Page 301------------

On the Count of

Three

"Garry, you know I have not touched the guitar for the last two years!"
"Tonight at Cafe La Puta, my brother.

We`ll send a het around. You`ll be

good, you`ll be rich." "Alright."

I hung up. A lightning struck the house opposite, for a while
illuminating my room. I caught my reflection in the mirror; there was a
trace of a smile on my face. Tonight, I

will be playing a guitar to accompany Allen Ginsberg. Tonight, at some
whore club, I`ll be the man to watch.

I went to the crapper. I was nervous, and when I`m nervous my bowels move
in a surprising rapidity. My swan was there, waiting to be used in a
lewd, inviting manner. I smoked a couple of cigarettes and did what was
necessary. I flushed and went back to my room. I had this apartment since
the divorce with Elisa, the only kindly remnant of memories I got left
from her. She was a swell lassie, but the problem was that she knew it,
or maybe it was the plastic ring I bought her for our wedding. Sincerely,
the only thing I know is that I was broke as a Theravada monk then.

One day something got into her, or more precisely, it was an Italian
carpet seller way downtown. Then it all went quick. She left the next day
she met him and took just a suitcase and two tickets to Egypt we had

125 pages the whole text..

--Page 302------------

On the Count of

Three

reserved for our summer holiday. Now, she sends occasional postcards from all over, but they never reveal where she is. Perhaps, if I were a bit better in names and Geography I could figure it out, but something ominously solicitous tells me not to do that. I love her still. She filled the horrendous emptiness that men experience after forty.

I went back into my room feeling lighter. The food I ate was as bad as the life I lived. On the dole for the last three years, I was becoming lazy and paranoid. I began visiting a shrink. The smallish guy with next to no hair and his Ph.D. pinned beside his patient`s sofa, compelling you to watch it all the time you were seated, was my vent of emotional frustrations I kept simmering under my bonnet. He gave me pills and I swallowed them. I did not care what they were for. I never asked. I just obeyed and swallowed. Perhaps that was my syndrome, a lack of maternal care in my early childhood. Now, mister Hankerson was my Mama, and perhaps even more; he was my spiritual guru, teaching me to take what was given.

I looked out my window. The rain was dwindling into a drizzle. Perhaps my lucky star was just coming up somewhere in the misty evening, making the whole affair of my life more bearable. I switched on a little lamp next to my sofa with a jerk of the switch-cord. The clock on a wall was showing six. I looked around the room. It was a dirty mess, but there among all the books and dirty clothes, yes, just in that little corner, there it was. My guitar.

I have not tuned that thing for a hundred years. I thought the

125 pages the whole text..

--Page 303------------

364

On the Count of

Three

strings might be too old and they would snap. They did not. I played a couple of cords. The wooden beauty cost me a fortune back in the day, and a few ulcers from starvation and sleep- deprived nights when working graveyard shifts to save a thousand bucks.

Before I bought this maple cat I was playing a guitar Evelyn gave me to my birthday. I was thirty then and earning most of my living by teaching kids.

Elisa wasn`t on the scene yet, and I was part of a local music recording studio called something like Local Metropolitan Orchestra. People in the neighborhood liked me

and because I was playing in the LMO, I was something. Children greeted me warmly in the streets and old men were taking off their hats. That is, before I was chucked out of the business for performing while riding a wave, as we say in California. No one cared that it was the best performance they ever heared. The thing I had done was hellish and the sixties then were gone with the wind. No one no longer thought of high performance and performing high as two necessities which should not be separated.

I started drinking and grew a beard. Now, children were scared of me and hats stayed firmly on people`s heads. I slouched my way through the streets like a stinking shadow. I lost the orchestra and began losing my students and my hair. Then Elisa appeared. She took me out of the gutter. She bought me a fish- bone soup in a Chinese restaurant. Then, I looked into her eyes and saw a deep blue sea of compassion and love. Before we left the bistro I stole a blue ashtray. Guess why?

Elisa took me home into a stucco house on Manhattan in the world of Not Much. I had three months of unpaid rent and after I recovered enough strength, on a sunny day in April, I went back to my place, took a handful of my belongings and left the room forever with a shaky wave of my hand and the-rest-is-yours farewell to my landlord. He remained positioned in the doorway like a stupid jerk, smiling after me till I turned the corner. I think his smile must have faded once he reentered my room and realized that the whole lot I had left him did not amount to a beaver`s shit.

Elisa Brandon, which was her full name, worked as a waitress

125 pages the whole text..

--Page 304------------

On the Count of

Three

in a bar not far away from where we lived together. She worked nightshift
so we fucked only on weekends. Her boss was a fat man from Alabama who
shaved his head to cover for bold patches.

He was of the good sort and I often came to that bar in the evening to
have a chin-wag and watch my lady serve. Then I would leave at two or
three in the morning drunk as a Russian and crawl my way home where I
would slump into a big armchair and sleep till dawn. Elisa moved like a
mouse, but her steps in the room always woke me up. She went to sleep and
I played on the balcony. That is, at least, during the summer. I`ve
always tended to remember the better parts of our relationship. After my
sacking for drug abuse during a performance, I could not find a real job.
In fact, I could not find nothing. I was so broke that it sometimes
became hilarious. I remember one episode when Elisa and me had our first
year anniversary. I went to the local cemetery on Gilmour`s boulevard and
stole a bunch of false flowers from a grave of some Mrs. Wilkinson. I
gave it to Elisa that evening. She smelled the petroleum-tinged roses and
said:

"Where the hell did you get that?" "Exactly there."

She trashed those flowers the next morning, but I remember having

nightmares

about the cemetery

and Mrs. Wilkinson

visiting me for about a week after.

125 pages the whole text..

--Page 305------------

366

On the Count of

Three

The

clock was showing half past six. I put the guitar back into its leather
case. It was time to go. Café La Puta sounded

familiar to my ears and looking into a small map of Lower Manhattan

in the world of Not Much, I discovered that the slut bar was

right

where I thought it was. I put on my Jerry leather coat and a woolen cap
with a big fluffy ball on top, a perfect outfit to August weather. I
walked into the drizzle. I was

walking

down the

125 pages the whole text..

---Page 306------------

On the Count of

Three

street, my guitar case in right hand. The rain was mild now and the street lamps were glaring into the gathering

darkness. I turned the corner and steered for the subway. I had to get over the river

somehow. I was

bouncing

down the steps into the lobby of the

underworld

of Not Much, when I noticed an old vagrant waiting

down on the platform. He saw me. I saw him. I dunno who was first. Anyway, it is an

interesting

phenomena, most

probably

correspondi

125 pages the whole text..

--Page 307------------

On the Count of

Three

ng with my darker past, that either I attract

beggars or they attract me. I have not figured that one out, yet. Anyway, this one went for me, alright.

"Got a dime for an old poor man?" "Got a rhyme about an airplane?" "Ah, mister is a joker."

"Get off my back you old trash."

He walked off. For a while I was feeling lousy that I did not give him anything, but then I remembered an old tune by Leonard Cohen. When they said repent, repent, I wonder what they meant… The car came into the tunnel.

Thinking that Mr. Cohen meant by his verse probably a completely different thing, I entered the car. It was half empty. I seated myself next to the window and watched the station disappear into black nothingness.

I held my guitar in both hands in front of me. I loved these moments. It had been a while since I played with anyone, and accompanying poets in their mad and ludicrous thoughts was a great fun. I began to smile at the passengers.

The door opened and I stepped out of the mole whistling The Summertime by Gershwin. Outside it was anything but summer. The night had fallen and the streets were like burning mirrors. I stopped on the corner to light a cigarette. A whore approached me.

"Wanna go for a drink?" "Your skirt`s too pink". "What about a place to eat?" "Now, look at your dirty feet". "But…"

"Shut up, ya slut!"

125 pages the whole text..

--Page 308------------

On the Count of

Three

She walked off, leaving a trace of cheap perfume. I was walking down the
streets again, my map of Lower Manhattan in my hand. I presume I must
have looked like a tourist, because when I stopped again to look at the
folded picture of New York, someone spoke.

"Are you lost?" "A bit…"

The man had no legs and with an amazing sense of balance he dangled in
front of me on two

crutches. He smelled of liquor and toothpaste. I thought that funny and
so I said:

"Are you a beggar or what?" "I am, you

snot."

"How do ya walk on those things?" "I have to,

it`s nothing amazing."

We parted.

I was on my way again. The traffic was mild and I kept my pace steady.
Thoughts were rushing into my mind. Things of the past. Evelyn once told
me how her Navaho father was gifted by gods and could do incredible
things. The story goes that when this Apache, or whatever, was hunting in
the fields of Not Much, he spotted his pray 50 miles off. Then he rode
his horse named Popo for two days in order to get it and terminate the
poor beast`s life. I was getting lost at last, and the rain hardened
again. Evelyn! Evelyn! Where`s my Navaho instinct when I need it? Then I
remembered, drenched as I was in that silly coat, an old mantra that
Evelyn taught me on an idle evening while we were both making roses out
of fag-ends sprang into my mind. The mantra ran like this:

Pi Pi

Apple pie

Give me sight of distant sky

I was mumbling under my breath. The street widened, lights of the
lampposts merged into a huge white-out. I walked on, aware only of my
ever increasing palpitation. I looked down on my feet and realized that
they disappeared. My physical imprint upon the world of Not Much has
melted away… Pi, Pi, Apple pie… I took flight. I was a beast of wing. I
rose to the height and saw a burning ring in the distance. I realized
that it was the sun, the only object on the horizon of a big, white
nothing. I was drawn to that luminous circle and felt no emotion except a
complete calm. As I drew nearer though, I felt the temperature rising. I
was sweating profusely and tasted salt in my mouth. Soon, it became
unbearable. I was getting right through that big, red smear. It was like
a second birth. Arrhhhh… Then quiet.

125 pages the whole text..

--Page 309------------

On the Count of

Three

I pulled at the handle of Café La Puta. What opened to my eyes was a long rectangular place full of many small tables with old chairs, all disappearing into a candlelit, cozy darkness on the left. On the right there was a small bar with a coffee machine and a couple of tap-beer handles. There were paintings on the walls, and on small easels scattered around the corners. Besides that, no one in sight. I always liked these wanna-be Art Nouveau dens, and some of the pictures looked interesting. I paused before one. It was an abstract painting full of lurid colours. It looked like a work of some spastic child who was kindly allowed to mess around with paints.

"A copy of Jackson Pollock. Amazing, ain`t it?" "Jeevs, ya scared me shitless, half-wit"

"Apologies, I work here. Would ye like something?" "I play here tonight. Fix me a drink."

"Ah, I see mister is also a poet,

for he speaks in rhyme."

"I`ve never written a single sonnet. Give me a break, I travelled in time".

The waiter left. I stared at the painting. Mr. Bollock, or whatever his name was, must have been rather off the wall to paint this way. The smears on the canvas, all red and purple, were designed into concentric circles. I pondered on. Nah, I`ve never understood art, and besides, who was he to know my little Navaho mantra with such a name. I seated myself, waiting.

125 pages the whole text..

--Page 310------------

On the Count of

Three

It was like waiting for your wife to put on her make-up. Unbearable. The guitar case was propped against the wall on my right. I was facing the painting, the bar was a long way off. I stared at the lighted candle placed in the middle of my rickety table. The flame mesmerised me.

It was my habit in the past to come to venues earlier. This time, it seemed, I was very early, or was it perhaps that Garry was late? What sort of a bartender was this lisping, spotty imbecile to take an age and a half to fix me a lemonade? Who was Jackson Bollock? Such and such questions were leaking into my mind, dripping like the water from my Jerry leather coat mercilessly flung over the back of my chair.

"Here`s your drink, I made it scotch." "You took eternity, watch!"

I showed him my wrist watch. It didn`t work. The waiter disappeared. I stared into the flame. It seemed as if the time has stopped, or was I getting paranoid from the time trip? Maybe there was some sort of jet-lag phenomenon? I rummaged in my leather coat pocket and took out a small

125 pages the whole text..

--Page 311------------

On the Count of

Three

bottle. It contained my prescribed pills. I poured them on the table. There were six of them in all, and each of a different colour. I put them one next to the other until I made a row. A rainbow, I thought, and looked up at the painting. This is a fucking art, Bollock! I crouched over the table and licked the pills up like a hungry dog. I drank them down with the scotch, leaned back and lighted a cigarette. The door burst open and Garry swaggered in, flanked with two whores. He was dressed in a corduroy suit and pink trousers. He pulled down his beret of green and flung it on the table in front of me. His flamboyant arrival had style.

"This is Linda and this is Sharon, from Maine." "To me, they look the same."

"Because ya have not slept with them, Brother." "I`d rather not. I have another."

"You mean little Lisa, I thought that`s an old flame?" "Better

a spark than two sluts from Maine."

"Don`t be so sarcastic and icy."

"Sorry, pal. I`m just tired of bureaucracy."

125 pages the whole text..

---Page 312------------

On the Count of

Three

They all seated themselves. For a while no one spoke, then Linda started to giggle. I ordered another scotch. Others also took orders. I finally began to feel like time was

passing again. Sharon kept on giving me the eye. I pretended I was blind to that, until Garry began to notice. The drinks came and the waiter broke the pathetic moment.

"How are you today mister Garry?" "Much better

now that I have my sherry."

"What`s new in the field of poetry if I may ask?" "A falling

twilight, my friend. Dusk!"

I gave Sharon the eye.

The place slowly started filling up. I was on the look-out for who-knows-what. There were many familiar faces and I would have sworn I saw Brad Pite. People were drinking and talking loud. Many voices carried western accents and it was obvious that they came to hear the star of the evening. Mister Iowa Gillette. Yes, that was his name, and it was him that I was waiting for. To hell with this bar and all these drunken snobs, there was a man with a real name.

The door burst open and someone came in. I heard the howl of wind from the outside. It seemed like the weather worsened again. Suddenly all the people fell silent and turned their heads towards the entrance. A dog on the floor stopped masturbating.

Gillette came on the scene. He was smallish but he seemed tall. He was dressed in white silk and his long black hair was braided. This man never smiled and never ever had anyone seen him eating. He came to our table, deep grey eyes piercing through our souls.

125 pages the whole text..

--Page 313------------

375

On the Count of

Three

Before Mr. Gillette`s glamorous appearance, I managed to get some information on his persona from Garry. The man grew up in a ghetto. His Mama found him in a trashcan. A little white child in an ape town, he was raised like a nigger. He stole when he was six and killed when fourteen. He grew up to be a man to fear. When he was twenty he got married. One day, he came home to find his elder stepbrother doing it to his wife. He killed him with a shaver and managed to get divorced before the police came. He spent forty years in a high security jail repenting and writing poetry. Behind the bars he was nicknamed Gillette. It stuck. A Zen master in the guise of night, he came tonight to read, to confess, to preach. He accosted Garry.

"What a wondrous night for a rhyme."

"That`s right Gillette, I hope we`ll earn a dime" "And who`s this man, all wet and drunk." "Well, this is Bill - to be frank".

"What is he doing at the poets table?"

"He`ll accompany you tonight. Trust me, he is able."

I got up to show that I was still a bit sober. Gillette shook my hand, measuring me with those deep eyes I mentioned. His grip was like a cold metal vice. He stared at my wincing face, then released.

"Glad to meet you Bill." "Your stare holds me still." "I do it to everyone know"

"Has anyone ever done it to you?" "Tonight we shall fly together Billy." "We are not birds, don`t be silly." "Birds and men, they both eat rice." "Hmm, interesting. Why, even wise."

Garry introduced the reading, as usual. He gave a short lecture on a book that had been around for decades and finished up with a sincere blabber on unprotected sex and red wine. Then he read a couple of his poems while Sharon and Linda from Maine danced around him smashed

125 pages the whole text..

--Page 314------------

On the Count of

Three

with cocaine, performing some kind of primeval dance.

I was just about getting bored with all the faded glamour and forced clapping when Garry suddenly stopped. Sharon could barely walk and she laughed like a poisoned baboon. Then silence fell upon the room, only occasionally interrupted by Sharon`s uncontrollable giggles. People looked up from their drinks to see Gillette walking to a small pedestal in the middle.

The little podium had two chairs. Gillette seated himself on the left and unfolded a stack of crumpled paper. He put on his glasses and turned in my direction. Everyone was waiting. I got up. This was the moment I was waiting for. I reached for my guitar case and pulled out my cat. I must say I was pretty drunk by then, and I flaunted the guitar like an auction piece.

I dunno how I got to that chair next to Gillette, the important thing being, thought, hat I had been sitting again. No danger threatened now. I felt confident. I looked at him and he looked at me. He seemed pretty certain about himself. As hard as a rock, he had the eyes of a saint. We had no amplification, but we did not need it. Gillette`s voice was like a thunder. When he spoke, it was as if from a distance, but then the words settled down with a splittering thunder. I listened to the first line of his poem.

Nightingale at night flees from

shadows to the light…

I played like I never played before. My fingers barely touched the strings. I danced upon my cat like a horny chimpanzee. Gillette`s rhetoric ability was imposing. He was throwing rhyme after rhyme as if he were stabbing daggers. One, two, three… Slash, Staby, Slash… There was a break and people cheered. I lit up a cigarette. I never

125 pages the whole text..

---Page 315------------

On the Count of

Three

felt better in my life. Never felt a better response from my co-performer. Never felt myself better responding. It was like the wildest intercourse I ever had. Like my best night with a woman. The only difference being that this was Gillette, and I was not gay.

He turned to me.

"Can you play blues?"

"Do ye wanna tie your own noose?" "Just

do what I say."

"Alright, I`ll obey."

I played and Gillette at first recited, then howled to the rhythm. Four, five, six… Slash, Staby, Stab… Suddenly I understood. He was killing his brother again, stabing him to death with that stupid shaver. Suddenly I was him. He was repenting for what he did. I knew it. This was his penitence and his confession. He spoke to the audience about orchards filled with dripping fruits, he spoke about dawn-wet dandelions in the fields of beauty, but always, he was silently telling the story of death.

As I said, our minds united. We merged into one single unit. We were howling like wolves. He turned to me while I was playing a solo and barked something I could not comprehend.

"Birds arise from burning rice to their own device"

I wanted to respond, but I could not find a rhyme. Instead, we took flight. It was like the effect of the Apple Pie mantra, only it was spontaneous. We left our jerking bodies behind. We were soaring high. A mild applause like rain echoek in our wake.

We stood on a high terrace overlooking emptiness. A burning sun in the distance. I could not fold up my left wing. Gill helped me a lot with that. We were

125 pages the whole text..

---Page 316------------

On the Count of

Three

watching the stationary pool of colours. Then Gillette asked.

"Do you know Jackson Bollock" And I said. "Not Much"

„One day you´ll be like him, only different."

„Thank you Gillet." „You are welcome." Six Hundred Seconds

Of

Freedom

Bie

Benjamin Schmidt

Suffering is the fleetest animal that bears you to perfection.

(Mistre Eckhart)

125 pages the whole text..

---Page 317------------

On the Count of

Three

On a planet which was by the local inhabitants often gladly referred to as "Blue", even after all the rivers and seas became polluted and turned murky gray so that from outside it looked like a big ball of something ugly and not blue at all, there lived a man that should never have been born. At least, that was what he always replied if asked. To begin with, the man was a madman. At least, that was what he often said when no-one enquired. His name was Isaac Burton.

He was born on a day that no-one had been expecting anything like that to happen, except for his mother who patiently awaited his birth for nine months and seven days. He came into

125 pages the whole text..

--Page 318------------

On the Count of

Three

the world like a punctual train hurtling to the station. Exactly at 11:11 am. When he was washed, wrapped up in a towel and handed to his parents like an oversized Chicken Wrap, Mrs. Burton and her husband George couldn't hide their surprise. Young Isaac was an ugly child, with a big Jewish nose and hairy ears that rendered him the look of a rat, and although the first thing he actually did was smile, it was with difficulty that his mother did not avert her eyes for the lewd, toothless grin made her sick, and so it did ever after, although she never told anyone in the long, lonely years that followed.

Mrs. Burton`s husband was an Englishman and she knew that such person as he was wouldn`t last for long and so it happened that five years after Isaac`s uneventful arrival into the world he died a demented death of asphyxiation with his own vomit in his bed. Mrs. Burton always afterwards ascribed this unfortunate accident to the English food she was, as a doting and caring wife, obliged to cook for him. And yet her husband`s untimely departure left her bitter and with little will to carry on, so that at forty she looked like a walnut and at fifty like a shriveled raisin.

While Mrs. Burton was subconsciously working on her physical destruction, her son Isaac was growing to be a man. It was his mother`s decision, after Isaac`s teenage years were over, to move to a bigger city where he would apply to an university and, due to Isaac`s love for languages, study English linguistics and literature. Although Isaac`s love for English was great and after the fashion of his deceased father he delighted nearly in everything that belonged to it, he was not aware that his knowledge was not sufficient to attend the prestigious school his mother in ignorance had chosen for him. His only equipage for the entering exam were the scraps of conversations he overheard as a child (because after his father`s death his mother refused to utter a single English word) and a Concise Oxford Dictionary he carried like a Bible wherever he went.

And so it happened that he failed and after succumbing to a short psychological breakdown that was only a foreboding of what was to come in his later life, he was forced to study a private school notorious for its incompetent teachers and drug-abusing students who, as one newspaper once succinctly put it, stood a bigger chance of dying of overdosing than actually graduating. Of course, poor Mrs. Burton knew nothing of these rumors and Isaac did all he could to hide from her the twisted reality he lived in as he later tried to hide his codeine habit.

The school consisted of five separate ground-floor buildings scattered in a big garden which in springtime was actually quite a wonderful place to stroll around. Four of the Units were meant for the students (one Unit for each grade) with the fifth, slightly larger, being the residing place of the headmistress and most of the teachers.

The headmistress was in herself quite a natural phenomenon. A fat, toad-like creature of unidentifiable age with at least one degree under every heavy-lacquered nail which she grew

125 pages the whole text..

On the Count of

Three

so long that they started to twist. She was definitely someone you didn't want to meet and most of the students were unassailably assured of her insanity. Nevertheless, she was the only person in the whole institution that actually demanded any sort of authority.

It is no exaggeration that a lot of students, with their minds half eaten by years of heavy drugging, actually suffered from a panic fear whenever Mrs. Smegma (for that was her name) was rumored to have been sighted somewhere in the vicinity of their Unit. Such people were basically unable to meet her face to face, and if anything so unfortunate happened, in five minutes one could hear the bleating of an ambulance-siren and a trembling, catatonic figure would soon be loaded on a stretcher and flicked to the local hospital. Those who left the school in such a fashion usually never returned. They either ended up on drug-addiction programmes or were simply never seen again. Although to this there was an exception. His name was Chalíl Trotzky. One day he was forced to face Mrs. Smegma due to some complaints from his teachers about his frequent absences. He met Mrs. Smegma in her office while heavily tripping and when shortly afterwards the ambulance came, the doctor pronounced him dead on the spot. The incident was hushed up and the whole thing soon forgotten.

One day about two years later, one of Chalíl's former classmates received a letter. It said:

Dear Yarden,

Although you've probably thought I'm dead, I want to assure you I'm not. It will be about a year since I came to the Himalayas and found a shelter in a Buddhist ashram. The incident at school left me with many things to ponder on and here I found peace and space to think. I want to find beauty in ugliness and terror.

Love,

Chalíl

Yet, Chalíl wasn't the only person seeking to have his deep questions answered. It was in the first year just one day before the start of the half-term holiday. The winter was very mild that year and one could walk outside only with his jumper on. Actually the winter was so warm that no casualties were reported among the students which was something deserving a celebration. It was quite usual that a completely drunk and loaded student, after the school was over, would forget the way out of the garden and then through exhaustion go seeking a refuge under trees so that he or she could be sheltered from the possible inclemency of the weather. Nevertheless, winter temperatures at night could go down as much as twenty degrees below. At such times it was quite frequent that in the morning with the frost still laying on the grass and crows perched on the naked branches, cowing in whispers as if they didn't want to disturb the brooding stillness, blue, frozen bodies could be found in the bushes in positions showing great death agonies.

But as had been said, it wasn't the case this winter and Isaac walking
out of his Unit when the school was finally over could even feel a sort
of lukewarm breeze caressing his longish dark hair. After half a year at
school he already felt there quite at home. All his fears and
reprehensions were gone and his psychological strength was growing day by
day as were his

125 pages the whole text..

--Page 320------------

On the Count of

Three

all the more frequent experiments with codeine. Isaac had been a strong smoker since he was fourteen, but exercised every morning and although through the lack of bodily fat which lent him a rather skinny and ascetic figure, he possessed strength that equaled that of his even much taller schoolmates, for Isaac was of a small stature. Isaac`s small growth was often a big source of his discomposure mainly in the presence of women, and because women had always been one of his fundamental interests, he was often troubled. It wasn't the case now, though, and as he walked out of the school precincts he decided to visit an old friend of his whom he new from the time he was attending his high school. His name was Jacob. Isaac and Jacob were inseparable and now that Isaac lived in the same city he visited his friend at least once every weekend. Jacob`s father Tamir was a man of great learning and something of a philosopher, although in his daily life he worked as a network supervisor for a local airport. His job was to control the airport computer network which enabled airplanes to communicate with each other and many other things beside, which when Tamir started to expound on, Isaac felt himself treading on uncertain ground because he couldn't make sense of the profusion of verbal intricacies which his vocabulary so abounded with, and later on he arrived on the conclusion that Tamir`s job must have been something of a philosophy too, or at least so Isaac thought because he had no idea what philosophy actually was.

Jacob was a good guitar player and Isaac played a bit of saxophone and so their favorite pastime was to sit and play together. Isaac then thought himself a poet which inclination was later to grow, and he often stopped blowing his sax to add a couple of verses:

The birch tree, she grew strong

Good people among,

The birch tree, she grew fast In the soil of red, dry dust.

But in the fury of vomiting fakirs She withered at last.

That day, after Isaac and Jacob enjoyed their instrumental pranks, they went to the kitchen to have an afternoon coffee. Jacob`s father put the kettle on and, after asking Isaac about his progress at school, embarked on a long monologue about the purpose of life and other necessities constituting a proper philosophical eulogy. But this time something in Isaac gave way and not only that he began to understand what Tamir was droning about but actually began to be interested. His attention was drawn mainly to Tamir`s narrative about mysticism and people who after long years of austere life and meditation were able to walk on the water or form clouds into funny shapes at their will. He was particularly interested when Tamir disclosed to the two listeners that, according to some eastern philosophies, there supposedly many dimensions in the universe occupied by creatures both evil and angelic, and that through the cultivation of the mind one can actually visit them. When Jacob`s father finished it was late at night, but Isaac wasn`t tired. He was buzzing with excitement he had never felt before and it seemed to him as if the air itself was charged

125 pages the whole text..

--Page 321------------

On the Count of

Three

with an energy of something mysterious. He asked a couple of questions on the topic and when he showed interest of knowing more on the particular theme of meditation, Tamir went to his library that consisted of a big wardrobe packed to the top with books of all sorts and carefully, with the tenderness of a child, picked out a book as if he exactly new what Isaac needed. Isaac, not knowing what exactly he needed took the book and read the title. The Straight Path by Raziel Haran.

His steps seemed unusually light to him when he walked home that night. Yet, in the morning Isaac woke up feeling lousy and when he went downstairs for breakfast his spirits took on an even stranger twist. He looked at his mother sitting at the kitchen table and it was as if he hadn't looked at her for the last fifteen years. She was a corpse. Her cheeks strangely puffed up with the excess of sleep on an otherwise lean face were spackled with yesterday`s make-up. His mother worked as an accountant for a small firm dealing with perfumes. She owned the job to the kindness of her neighbour who took pity on her after her husband`s unexpected departure and although she was now taking most of her work home, she was hardly coping. Mrs. Burton seemed to wage a bitter war against something that was slowly working on her complete destruction, and she seemed to be losing. After he had eaten, Isaac went to hide to his room and finding the book he borrowed the previous day he started to read.

The book was a narrative of a mystic that lived roughly a century ago and Isaac got immediately absorbed into the seemingly weird and improbable practices of this unknown person and when he got to the chapter describing something that Jacob`s father mentioned the previous night as the cultivation of the mind, he put down the book and following the description of a picture he crossed his legs and, resting each hand on his knees, for about twenty minutes pretended to be meditating. In this fashion, through the book by Raziel Haran and many other books that followed, a seed was sown in Isaac`s barren fields of his life that was later to be a source of both his happiness and near undoing.

The school days went fast and Isaac soon found himself moving into another Unit which represented his ascendance into the second grade, yet Isaac was sulking. His knowledge of English for which he came to study The Wilbour`s Institute of Languages was no greater then when he started. Yet, through his depression he couldn't subdue the feeling that it wasn't utterly his fault. Actually, it was not his fault whatsoever. The school`s notoriety as one of the worst learning institutions in the country seemed to have been not only a rumor, nor a simple blatant fact, but rather something that must have had taken years of painstaking struggle on the part of the school staff. The struggle of a crippled whore, as one of Isaac`s friends later aptly remarked, was visible everywhere.

Not only that most of the teachers had the habit of coming late for their lessons but some often would not turn up at all. There was a Polish teacher of philosophy named Horatz Soltz, who although for his laid-backish nature was quite favorite among the students once didn`t

125 pages the whole text..

On the Count of

Three

show his face for three whole months. One day, just when the spring could be felt in its full, the door to the classroom opened and Horatz strolled in sporting a grimy, old shirt that had but one button left and which, with obscene nakedness, showed his fondness for good food and drink. He jumped behind his desk like a wild boar and with his keen, dark eyes started surveying the classroom. The attendance, as you may expect, was rather low that day since nobody expected to see Mr. Soltz ever again. There were three people sitting by the glassed wall through which you could see into the garden and in the back, a girl trying to read Koran in English who probably didn't even know what she was doing there. "So Omar, you thought you was gonna escape the oral?!" He barked. "Well, you were quite right to think that," he demurred.

"In the last couple months I've seen enough stuff to last me a lifetime, so don't you guys

give me no crap about Joyce. I want to hear how you all are. Understood?"

Nobody moved.

Such was life at Wilbour's and Isaac soon began to realize that unless he started working on himself he would never become anything worth being, whatever that might have been. So, one day he went to a bookstore and bought an English dictionary of such thickness that it filled his whole schoolbag and set to work. At first just memorising words he did not yet know, and when he thought he knew enough, Isaac moved to translating word by word giants like Shakespeare and Yeats. He also stopped minding if the particular author was British or American. He read Charles Boudelaire, Paul Verlaine and all the cursed lot. He read Salman Rushdie and Roald Dhal. He read a lot. It was also at this time after perusing Lawrence Ferlingetthi's poems and William Burroughs' Naked Lunch that he discovered in himself a great relish for swearwords and things obscene which subconsciously made up for his torn and meagre love-life. Isaac began to grow both intellectually and spiritually. Every fortnight he would visit his friend Jacob with the twisted purpose of talking to his father. Tamir would gladly lend him books dealing with the metaphysical, mystical and who knows what other esoteric teachings that his ample library abounded with. Isaac would read, and then read again, and then seat himself in the shelter of his room and meditate while his mother, with her early senility kicking in could be heard sobbing downstairs over the pitiful creature that she had become. Her raisin phase was now long in progress and Mrs. Burton could hardly be said to be sane anymore. Yet somehow, even through her rapid psychological decline, she found in herself something that held her firm among the living. It was love for Isaac that was saving her from complete desolation, and yet it was a love that her son in his early adolescence could hardly discern, and so they often quarrelled.

◇

125 pages the whole text..

--Page 323------------

On the Count of

Three

Isaac`s friend Jacob was studying sociology with the emphasize on the care for the disabled, and as Isaac`s interest in the spiritual was day by day more visible, Jacob`s intrigue began to grow as well. And so it was that on a Saturday evening one might find them walking towards a rather obscure part of the city, and after zig-zagging through a small maze of streets, enter a place that was under the general knowledge simply known as the Tea House. In this place seated on a carpeted floor, drinking oriental tea and smoking hookahs filled with an aromatic tobacco they would exchange their little views on things both mundane and pious, and such practice they proudly called a philosophical discourse.

Jacob`s opinions were often forcibly declined by Isaac who firmly stood his ground although he himself wasn`t often exactly sure where the ground actually was, or what the hack he was talking about, or even why he was shouting all over the place: "You can`t mix up subliminal experiences with open revelations". Yet, what the phrase actually meant was anybody`s guess.

Isaac`s impulsiveness, as he would later come to see, was due to an envy that he

unknowingly fostered towards Jacob. It was because Isaac had very few friends and even with those he didn't seem to have much in common, while Jacob always gladly and heartily talked about his mates from school and what a great time they had partying during the week.

Isaac felt helpless. He had taken upon himself the burthen of self education while obligatorily having to visit a crippled whore, for he wouldn`t now talk of his school in any other way, and because of his days filled with hard work, he basically didn't have time for social life.

Yet, one Saturday, about the end of his second year, he received a phone call from Jacob. His voice was full of glee and he invited Isaac to join him and a couple of his schoolmates to the Tea House. Isaac was just then slowly coming to terms with his physical appearance and when he was made to understand that a schoolmate didn't necessarily imply a boy, he said he would definitely come.

Isaac arrived at the spot at quarter to seven which was approximately the appointed time, and guided through the labyrinthine space of the Tea House by Jacob`s ringing laughter that he knew so well, he tentatively led his way to the back where the tea-room was divided into separate compartments for the sake of privacy of varying nature.

He pulled back the curtain and found himself in a conflagration of loud voices, smoke and people he didn`t know. "Isaac, I am here!" Shouted Jacob`s disembodied voice that, as Isaac`s blinking eyes became adjusted to the dark space, soon took on a physical form. All in all, there were about five or six people roughly of his age and everyone, when he entered, looked up to him with merry eyes. Isaac seated himself in the corner and Jacob started the introducing. Isaac`s memory for names had always lacked a flair and he had soon forgotten

125 pages the whole text..

On the Count of

Three

how to call the shadowy silhouettes that made appearance one after
another in front of him. Yet, there was a name he remembered ever after.
She was sitting in an opposite corner. Her hair was hay-blond, her eyes
blue, and her smile by which she immediately encharmed Isaac, when she
locked him in her stare, was enigmatic and yet secretly telling of beauty
and passion. "This is Shirel!" shouted Jacob through the loudness of the
room. "We sit together at the same desk at school!" But Isaac paid little
attention to what was Jacob saying. He suddenly felt himself in love with
that girl and words like hump, bong, and jazz were all on his mind.

He spoke little that evening and all the time he was striving not to meet
Shirel`s eyes again. At least, not in the grippling fashion in which
their looks met for the first time, because he was afraid to see what her
sea-blue eyes so blatantly showed. He was scared of seeing that
understanding look that talked to him of all the degraded passions that
he secretly dreamed about. And yet, through all these manifestations of
debasement she seemed to him like an angel of compassion.

The party soon began to disintegrate and Jacob and Isaac later found
themselves in the streets with a slight drizzle wetting their clothes. On
the way home Isaac strove to glean out some information on Shirel
Thorsly.

While Isaac was midway to his nineteenth birthday, Shirel supposedly
boasted with the sober age of twenty-four. She had allegedly studied some
obscure school which after leaving turned you into a waitress in a shabby
bar, at best. She cancelled her studies halfway through because she
wanted to become a social worker and, according to Jacob, she was very
intelligent, although Jacob, in Isaac`s eyes, didn`t seem to notice the
god-like nature which she so vividly emanated. Nevertheless, before they
parted that evening they agreed to meet in the same crowd and manner next
week, and it might be said that Isaac`s steps seemed unusually light to
him when he walked home that night.

And so it was that Isaac`s life changed. If to the better was hard to
tell because he now lived from weekend to weekend with the prospect of
seeing his enigmatic mistress on Saturday evenings. During the week he
was a walking corpse like his mother was, which sudden change led him to
think that his mother must have been either desperately in love with
someone, or at least that she must have dearly loved his father for whom
she was now turning into a withered scarecrow. Of course he wasn`t far
from the truth, but he could be hardly bothered as he was now existing
for two things only. Shirel, and codeine.

And so it was that their relationship and mutual affection was growing
week by week, but there was something black and spoiled like bad food,
that slowly and innocuously crept up into Isaac`s brain and which
appeared fully in light on the summer holiday.

125 pages the whole text..

---Page 325------------

On the Count of

Three

◊

Isaac was enjoying the summer at full blast and he and Jacob were often to be found together doing what they started to call practicing, because they now thought of themselves as a band.

They would play together from early afternoon till late at night, and then, after a hard day`s slog, sit in the kitchen of Jacob`s parents` house while everybody was asleep, and in the blue haze of cigarette and opium smoke hold long conversations which Isaac would from time to time intercept with a poem of his own making:

What a blight on a stormy night, When I found my ship near sinking. From afar a saw a light

Through the murky clouds but limping. Thou guided me to a serene sea, Thou gavest me a courage.

And soon I felt the homely warmth Of thy harbourage.

"You pervert," cried Jacob, laughing. "She really twists your brain, that girl does." But Isaac could not help himself. His mind was drunk by Shirel`s inebrious influence. Nevertheless, what came afterwards was even headier.

"You know that Shirel is going away next week?" Said Jacob after a moment. "Really, where?" inquired Isaac with a lump in his throat.

"She was asked by the family of her schoolmate to look after their small ranch for a fortnight." answered Jacob. "They go rafting or something, and they have a couple of horses to take care of." He finished.

Isaac was speechless, and when Jacob saw his dejection he started laughing again, and said: "You fool, I mean, it`s your opportunity! Don`t you see?" And Isaac saw and his eyes were glazed with lust. He understood Jacob`s twisted stratagem and he silently agreed to what didn`t have to be said anymore. "You go and ask her if she would like to take you along"! Helped Jacob, not understanding that Isaac already understood.

"Fourteen days of sexual merriment," said Jacob, venting his thoughts. "I must confess I envy you. "

◊

125 pages the whole text..

--Page 326------------

On the Count of

Three

Isaac came home feeling elated. He went straight into his room and locked himself up. "This will require more than just a phone call." He thought. "And there`s no place for mere luck here. I need a spiritual backing". He seated himself on the floor and crossed his legs. He was pretty good now at doing half-lotus posture which he saw in so many pictures of practicians, and when he was sitting for about five minutes concentrating on his lousy breath, because he forgot to brush his teeth, he dedicated his mind to metta , a type of Buddhist meditation also known as the meditation of unconditional love. Yet, this time he was the one to call the shots, and he conditioned his concentration on one person only, Shirel.

He was sitting, breathing heavily, and with his eyes set on some undefined spot somewhere in front of him, Isaac tried to muster as many positive thoughts as he could possibly think of. But something down below in the vicinity of his groin was disturbing his spotless thinking. In half an hour Isaac was lying in bed felling depressed and sort of empty from his waist down. "Damn!" he thought to himself and fell asleep.

The morning broke out and the first rays of the sun found Isaac franticly pacing his room. He had been chain-smoking from five o`clock and his tar-stained fingers were trembling. "I`ll call her at ten." He said to himself and slumped on his bed. The fingers of the clock seemed slow and the time was dragging on like a sermon. The air was charged with an infinite impatience that was hard to bear. At half past nine Isaac picked up the receiver and started dialing. Mrs. Thorsly picked up the holler on the other side. "Is Shirel home?" Breathed Isaac, and when Mrs. Thorsly went to fetch her, Isaac fell into an apoplectic fit of coughing caused by the morning`s excessive cigarette abuse. Shirel reached for the telephone and for a bewildered moment listened to someone dying on the wire. "Shirel, is that you?" blurted Isaac when the coughing stopped. "Hi, Isaac. Why are you calling me so early?" She enquired.

Isaac was about to relapse into another fit but he sustained himself and continued. "I was thinking if you`d like to take me along to .." he nearly said the sexual merriment "Ah, well, ...the ranch?" He spat out with exhaustion. There was a momentary silence. Then Shirel spoke. "But of course, I myself wanted to ask you if you`d only let me." Isaac hung up. He was too happy to speak and anyway, there was nothing to talk about anymore. The rest of their relationship should be only carnal satisfaction.

✧

Yet, Isaac did not know what an unfortunate episode he had brought upon himself and poor Shirel by going with her to what was later to become a walk through infernal fires. They set out on a bus going westwards on the fourteenth of July which took them most of the way

125 pages the whole text..

--Page 327------------

395

On the Count of

Three

to the ranch. The rest they walked. About six miles of a winding path running through meadows, two small villages, and a forest that Isaac took an immediate liking to. They spoke little on their way, for Isaac felt his words stuck in his throat and Shirel was talkative only when roused by a vigorous conversation. Isaac was unable to speak because his mind was preoccupied with numerous thoughts which origin and sense he could discern with difficulty, and when they passed through the forest of pine and beech and climbed a small steep to where the paddock of the ranch was beginning, he forced himself into: "Here we are, aren't we!"

They approached a high brick building of palatial dimensions which was to be their abode for the next fourteen days and both Shirel and Isaac stared in amazement at the winding stairs leading to the main door and two high columns bearing the weight of a spacious balcony on the second floor. The house was big. Yet it was just then, when staring at the opulence, that Isaac`s thoughts took on a more definite form and for the first time he was aware of the disunity of his mind. When he glanced at the sprawling mansion nestled in the little grove of beech trees and than looked at Shirel, no taller than he yet beautiful, and what seemed to matter more now, simply a woman. When he thus looked at her he felt nothing but sexual desire, and at the same time he was conscious of the baseness and untimeliness of his thoughts, and he desperately tried to banish them. Shirel might well have remained an angel of compassion to the eyes of others, but Isaac now saw in her only what he saw in himself. It was a debased and twisted love that was to give little to the other.

They entered the house and found a corridor from which many rooms of uncertain purpose were running in all directions. After half an hour of scouring they found the living-room which also served as a bed-room. Shirel silently stepped in front of the bed and began dressing the pillows and sheets. Isaac stood close by, looking on. Yet, he could not help feeling a sudden disgust. He was angry at himself for his muteness and enraged by the thoughts that oppressed his mind. He left Shirel and went out onto the porch. The sun was falling behind the house and calming darkness was quickly setting in. He opened the case of his saxophone and started playing. Shirel soon appeared, and seating herself on a chair she listened to his maudlin bleating. She was a good listener and they spent a couple of hours in this rapturous trance which was later to be their only time of repose in which silent gregariousness they seemed to escape from one another and yet to be somehow firmly bound.

When Isaac finished it was late at night and Shirel got up. She beckoned to him and by this nonverbal gesture Isaac understood that it was time to go to bed. They walked in silence through the corridors of the house and Isaac couldn't help feeling like a lost child afraid even to whisper what he longed so much to cry out, as if not to disturb something in his

companion, or perhaps to rouse in himself something that he secretly feared. Yes, it was the evil part of him that Isaac was afraid of and what with the arrival to the house began to knock at the door of his consciousness. The kind of evil that everyone of us takes a share when we come to the world, yet this was something Isaac did not know and was completely

125 pages the whole text..

--Page 328------------

On the Count of

Three

unequipped to deal with.

They arrived to their common bed and silently slipped in, and through the
dark void that surrounded them Isaac could hear Shirel secretly sobbing.
He was desperate, but unable to move. Unable to avert the course the
events were taking. It was like taming a wild river and he felt
powerless. Shirel quickly fell asleep and Isaac for hours that stretched
in the vastness of the room into eternity listened to her mellow breath
and was confused.

The daylight opened Isaac's eyes and he found himself alone. The events
of the previous day seemed like a bad dream that he might have just
awakened from. He felt refreshed and the effulgent light coming into the
room through the high windows raised his spirits. He set out in search of
a kitchen which he had not had any previous knowledge about and after
blundering through many rooms that interconnected into a spectacular,
architectonical labyrinth, he found Shirel in a high-ceilinged room with
windows looking westwards, and because the place seemed to have all the
faculties of albeit a bit larger kitchen he concluded it for a place
where food was generally prepared.

Shirel was making tea and toast with butter and accosted Isaac in a
manner that showed that even she must have forgotten the previous day's
hardships and her night weeping. They had a short conversation over their
breakfast and Isaac felt that everything might not have been lost, that
is, if it weren't for his mind that was subconsciously working on his
slow plunge into insanity.

When the tea was nearly finished Isaac suddenly realized that with the
precious liquid gone the morning's little idyll would expire too, and he
was seized by worries that seemed to gnaw on his marrow and petrify his
tongue. He looked at Shirel and felt a sudden, deep loathing towards her,
yet for what reason he could not tell. Shirel, through her receptiveness
immediately saw in Isaac what he himself was only slowly beginning to
realize. He was losing his mind and losing it quick.

They were sitting for minutes that seemed unbearable, unable to meet each
others eyes, until Isaac no longer able to withstand the pressure looked
up and saw Shirel's scolding look that at the same time mirrored his
confusion. He got up and fled.

He fled to the forest that he took such delight in on their way to the
ranch. He spent hours walking among trees or sitting on a big stone slab
that he himself in his despair discovered. He now knew that something was
out of order and that he began losing grip on reality which when he
looked around him seemed nothing more than just a film-set in some
surreal parody. Yet there were his thoughts and they were more real then
he ever felt them before. Real in their oppressive strength but bizarre
in what they disclosed to him. His thoughts talked of horror which had no
object and he tried to run away. They talked of Shirel and her beauty and
he pondered on his returning. They talked of ugliness which he suddenly
saw wherever he looked and he stayed where he was, sulking and raving.

125 pages the whole text..

On the Count of

Three

The evening was coming and Isaac became tired. And with the weariness of
the body calmness came into his mind. Isaac suddenly felt foolish and
childish for his running away and he even began worrying about Shirel. He
slowly walked back and when he got to the place where the wooden planks
marked the beginning of the paddock he started to follow it. In a while
he discovered Shirel. She was standing on the other side of the fence,
pouring water into a big trough. All in all, there were four horses on
the ranch, and they were slowly coming towards her. Tentative at first,
then giving up to her gentle hand which stroked their heads that seemed
strangely unreal in the falling twilight. Isaac lingered till Shirel
finished her duty and they walked to the house together. They spent the
evening in their silent solidarity, with Isaac playing his saxophone
which mellow tones were being answered by neighing of the horses in the
unseen distance, while Shirel seated on a chair as the previous night
listened attentively and seemed at peace with herself. Then the bed-time
was come and they parted in their thoughts with each other because they
new what was coming. They slipped under their blankets and pretended that
their coldness and the space between them which to both seemed like an
intolerable abyss was something they could ignore or overlook, and yet in
their half-nakedness they felt pathetic and immature.

Days went by and if anything improved it was either stifled or overlooked
by what became worse. Their mornings were the only time when they were
able to talk to each other, but as time dragged on Isaac slowly began to
realize that he was unable even to meet Shirel face to face and they
would spend many days hiding from each other in the labyrinthine
passageways of the mansion.

After twelve nights spent at the ranch Isaac saw his sanity disintegrate
in front of his eyes like burning paper and Shelly for her fragility
didn`t seem to be better of. In the evenings they would both crawl out of
their hiding places as if some atrocious game was finally over and sit
over a pot of tea listening to the silence of the night. There began to
be something animal-like palpable in their behaviour in which strong
amity was soon exchanged for near hatred, and there was a certain
weakness or fatigue that rendered them powerless against the demon who
brooded over their tired souls.

And so it was that the last night before their departure came, but Isaac,
although deranged and exhausted, didn`t forget his purpose on the planet
as a representative of what little manhood there seemed to remain in him.
And so it happened that when he and Shirel slipped in bed, weary with the
nonsense of the day, Isaac understood the singularity of his chance of
ridding himself forever what had been oppressing him for so long. His
virginity. He spoke and his voice, rough and harsh by so many days of
disuse, sounded through the dark like a menace. "Shirel?" He growled.

"Yes?" Shelly answered tentatively from an undistinguished distance.

"I can`t see you. Where are you?" he enquired and, for the first time
realizing that he was speaking, he was

125 pages the whole text..

On the Count of

Three

seized by worries if she would repudiate.

"I`m here. Where are you?" She answered, her voice playful and expectant.

✿

Their journey home was the happiest event of the last fourteen days. They spoke little and if they did it was to discuss the bus timetable, or how long it would take the bloody bus driver to finally realize he was on a highway, and speed up. Isaac and Shelly couldn`t wait for the moment of their parting and when it came, it was as tepid and off-hand as if they were supposed to meet on the same spot in fifteen minutes. Nevertheless, it was many years after that they met again and were able to hold a normal conversation, although what happened at the ranch was a secret they didn`t even dare to share among themselves.

In the meantime, the summer holiday was slowly drawing to an end and Isaac was spending his last days of freedom with Jacob. When the school began it would be his third year at The Parkinson University, and the prospect of being half way through with his studies made his spirits soar. Yet little he new what an unexpected turn was his life to take after the holiday was over and that he wouldn`t have to meet with any of those wrecks who clamed to be his classmates ever again.

The evening came and it wasn`t as warm as the previous days. Tomorrow would be the first of September and Isaac could already feel the autumn coming. He liked autumn and the sort of weather when it wasn`t bloody hot. He packed his schoolbag and went to say good-night to his mother.

She was already in bed. Wrapped up in five thick blankets with just her head sticking out she wasn't dissimilar to a larva waiting for her final faze of turning into a butterfly, yet what she was turning into was much more digressed and appalling. Nevertheless, Mrs. Burton seemed to bear her fate bravely. She was reading Man`s Health and on her wrinkled face Isaac could discern the smile of a saint. "Night mom." He said, and not waiting for an answer he slowly closed the door.

The sun appeared that morning in the West. Quite a normal phenomenon which was later on to balance out the things that were completely out of order. Isaac was late for school, and the day being the first day in his third year he didn`t feel wholly satisfied with himself. He arrived at Parkinson`s with a two-hour delay and found a big crowd of drunken people in front of the main entrance.

Thinking that it was rather early for a party of such dimensions he squeezed himself through the gate and to his disconcertion found even more people inside. Isaac soon realized that something was amiss and that this wasn`t just a normal students` party. There were adult

125 pages the whole text..

--Page 331------------

On the Count of

Three

people, too, and they didn`t look like the teachers he new, nor like those he was yet to get to know. They looked conspicuously like the parents of the students. There was an old woman which, as he later realized, was a granny of his fellow student. She appeared out of nowhere, yet how she did it with that heavy cane and a strong limp was beyond Isaac`s currant worries. "Got a fag?" she asked him with a benign smile. Isaac was so flabbergasted that he even offered her a fire and gentlemanly lit up the cigarette in the old woman`s trembling mouth. Then, just when Isaac was at the height of his confusion, thinking that the whole world had perhaps gone nutty after all, his classmate Roona came to meet him and greeted him with a flourishing exuberance.

"Hey, Isaac! So, here we go after all," he said, and Isaac understood that he knew something he didn`t. "What in the name of all marvels is going on here?" cried Isaac at Roona through the raising commotion. "Don`t you know?" Roona said with a winning smile. "She`s gone!"

✧

Isaac was sitting on the grass under a big apple-tree. He and his five classmates were sharing a bottle of wine. The news was too bad to be true. Mrs. Smegma grabbed the stash and took to her heels. That was the end of the school and dreams shattered for Isaac of ever obtaining a degree. He was looking at the crowds of parents that had just cornered a few teachers who shared the same uncomprehending expressions and calling for answers till those were given, the staff were soon calling for mercy. The police soon came and, first of all, chased out a half dozen tramps who came crawling out of their summer hide-outs smelling alcohol and diversion.

Indeed, for some it was a cause for a celebration. Nevertheless, even more interesting was the disunity of the students` opinions on the actual situation. There were a lot of pupils who were truly stricken by the horridly blatant fact that Mrs. Smegma was gone and gone for ever. The majority, of course, didn`t give a flying walrus about what has happened because they had other, more pressing things on their minds which mainly constituted in how to transport fifteen beer-bottles in a school bag from a local shop which couldn`t possibly hold more than twelve. Nevertheless, there was a small group of students who actually claimed Mrs. Smegma`s sly run-away as the only smart thing she had probably ever done in her life.

Isaac did not know what to think. He felt strangely empty, yet what he learned from the many spiritual books he read was that emptiness was a positive sign of the aspirant being on a good way to his ultimate goal. And as his rage and sadness slowly began to melt into the background, to Isaac still sitting on the soft grass it seemed that perhaps things were supposed to be happening that way.

125 pages the whole text..

--Page 332------------

403

On the Count of

Three

And just there and then, amidst the commotion, and half-drunk with cheap wine, he received a revelation. It wasn`t as much about what he saw as what he felt. It began as a strange prickling in the stomach and then he felt as if something warm and unbearably light was spilled insight of him. It spread into every part and extremity of his body and he suddenly thought himself enlightened. He rose and felt tall among the circle of his drunken schoolmates. He was standing and his eyes were looking far, yet none of the onlookers knew that he was surreptitiously watching a woman outside the garden who was walking by in a mini-skirt. "Fickle is the harlot`s tender bosom. I`m leaving," He declared, and walked away.

◊

Isaac had always been told that adult life was much harder than the life of a teenager, and the marks of heavy use which his mother wore on her body like a stigma seemed to him to be a good enough proof. Nevertheless, after about three weeks of succumbing to sulking and heavy depressions which came like an unexpected storm after he walked out of the school precincts, Isaac came to the conclusion that if there was something he wished in his life it was never to meet the fate his mother seemed to be destined to. He bought himself books and then a couple more, and with the dedication of someone who has a wish he again dedicated his person to a persistent study of the English language. Yet there was something he soon began to feel like the prick of a needle. A needle that his mother herself started to jab him with in the quarrels that became his daily bread and butter. "You have to find a job." She would say. "Willy-nilly you fucking gotta." Nevertheless, it was a thing easier said than done. At least, that was what Isaac thought and he reproached his mother for her urgency in this matter, although he himself soon began to feel the tinge of idleness. And so it happened one day that Isaac was walking through the city center. The summer was slowly receding and giving way to more bearable conditions for a creature of his measly dispositions. He was just walking by a big, opulent building that had been under heavy repairs for a couple of years and was known to the general public as The Municipal House, whose worn frescoes and stucco decorations dated back to the Art Nouveau period.

Isaac had always admired the building although his knowledge of styles and Fine Arts in general was meager. He approached the building and gazed in wonderment at the structure in front of him. Nevertheless, it was just shortly afterwards that something else attracted his attention. He was watching the workers and it was just then that he realized that they weren`t actually working on the building, but on something inside.

His curiosity gave a push to his legs and as he prompted his steps towards the main gate through which all the people in helmets and clothes soiled with dust seemed to be disappearing, it occurred to him how beautifully simple it must be to do this sort of work. Not simple in being trivial. But simple in being just hands and muscles. It was also then that Isaac understood a bit his indulgence in thoughts and how little attention he had paid to

125 pages the whole text..

On the Count of

Three

mundane things, and allowing himself another submergence into the world
of dreams, he took upon himself the idea of becoming a worker.

He passed under the high arches of the gate through which in times of old
marched the soldiers of the city, because the house before refurbishing
originally served as barracks, and what opened itself to his eyes was a
sight he remembered ever after. The whole courtyard of the old barracks
which in largeness nearly surpassed the main square he had just left
behind was all dug up. Deep trenches were delved into the ground
throughout the whole area and one could see people climbing ladders and
handing each other buckets of earth, and women standing on the surface of
this surreal, archeological mine were carefully examining the contents
with the strokes of their gentle hands. So Isaac immediately understood
that he had entered a world different to the one he knew so far. A world
of forgotten things waiting to be uplifted into the light of curious
eyes.

Awe stricken as he was Isaac missed to notice a sturdy gatekeeper who had
been all the time standing by. "What do you want here?!" He growled,
measuring his skinny, bent figure. "This place is out of bounds for
people who don`t belong here." But Isaac, averting his eyes from the
splendor of the excavation works, was already far ahead of him. He now
thought himself an archaeologist. "I work here." He said, and gave the
bulk of a man a long and piercing look that made the fat drunk step back.
"Go ahead then, sonny," stammered the gatekeeper. "No offence meant," and
he made an opulent gesture as a token that his passage was free. Isaac
pulled himself up to his full height and reaching the gateman nearly to
his shoulders he made a tentative step into an episode of his life that
was to be both joyous and harrowing.

In fact, it took fourteen days before all the necessary papers were
signed and Isaac was officially accepted into the archaeological dig. The
morning was heavy with anticipation when Isaac was choosing a proper
working gear from his wardrobe for his first day at work. He picked an
old pair of trousers and a warm jumper the brown color of autumning
leaves, for the summer was now long gone by and there were chilly days to
come.

He got to the city center at half past seven and passing again under the
tall arches of the main entrance he greeted the warden like one who
thinks himself high. The building of the old barracks, as Isaac soon
understood, was fully devoted to the ongoing excavations. He climbed to
the second floor and found a dressing room that served both for men and
women. He dressed himself slowly, yet not for fear of what was to come,
but out of diligence and dedication.

Isaac was an archaeologist and he felt his duty to behave so. When he got
dressed and allowed himself a surreptitious look at a girl that was
chastely trying to pull a pair of trousers over her fine legs and scarlet
knickers, he walked out and descended the broad stairs leading into the
courtyard. The morning was chilly and a slight, morning mist was still
crawling in the deep trenches which the digging-ground seemed ridden with
like some exclusive Swiss cheese.

Isaac was instructed to take a pick and a shovel and when he loaded both instruments on a

125 pages the whole text..

--Page 334------------

On the Count of

Three

wheel-barrow he set off across the Emmentalian surface to his prescribed
destination. He was to work with two other boys approximately of his age,
and when he arrived to the hole in the ground that was to become his home
till the beginning of the winter, he gazed down and for a while savoured
the ephemeral moment of his seeming triumph, for what he saw in the pit
was more than anybody else, except for rare exceptions, ever noticed. He
saw happiness.

✧

The dig proved hard, because all the fine work of sieving sand and
cleaning unearthed pottery was reserved for women, and thus Isaac found
himself slogging the pick from six to eight hours a day. He wasn`t used
to such labour and the first three weeks suffered greatly from a severe
pain of all his extremities. Yet he persevered and his sore limps had
grown accustomed to the constant shoveling of earth and picking up heavy
stones.

Isaac grew strong. Where there had been nothing but skin and tendon were
now muscles, and his hands were callused and hard to the touch. There
were many people working at he excavation works and Isaac soon got to
know them. Some were young and pleasing to talk to, some were older with
lots of experience, and some were conceited pricks. There was a man Isaac
loathed in particular. That man was about fifty years of age,
approximately the age his father would have been if that poor soul was
still alive. He had a prematurely white and thinning hair which he had
the habit of tightening at the back into what could be described only as
a very small pigtail, or rat-tail to be exact. The man wore a goatee the
colour of his hair which size and shape along with the appearance of the
rest of his face rendered him the look of a Chinese poet of old times.
Yet the man was anything but a poet. In fact, the man could barely talk,
and if he did, it was nothing Isaac would really miss not to hear. "You
are young," he would say to Isaac. "You were still in mummy`s tummy when
I was chasing after girls and pouring vodka up their cunts! Ha,ha."

Isaac worked from eight in the morning till four in the afternoon, and
when he came home he read books of all sorts. He studied till deep at
night the English versions of the secret teachings of Maister Eckhart and
Shopenhauer, then got up early in the morning and went to work.
Nevertheless, although most of the books he read dealt with things
spiritual, Isaac through his vocation soon became interested in geology
and other scientific disciplines prodding deep into the origin of the
human and the beast. It didn`t last long and he developed a rhythm in
what he was doing and he soon began to reap the fruit of his ardent
devotion. He felt the happiest in his life and everything he touched, or
did, seemed to be steeped in something he could refer to only as holy.
Yet holiness that thinks itself holy smells and as time drew on the fruit
he had reaped began to rot.

125 pages the whole text..

--Page 335-------------

408

On the Count of

Three

December came and with the Christmas at hand came snow. Yet Isaac was
working as hard as ever. There was a lunch-break at twelve o`clock every
day and Isaac would stand at a high table in the canteen, sheltering a
cup of coffee with both his hands and with an unceasing vigour exchange
information about the newest finds. He felt himself an expert on
dinosaurs and although there were none to be found in the whole country,
after a coffee and a couple shots of cheap Brandy he would heartily start
to expound on his far-fetched theories on the eating habits and moving-
patterns of these prehistoric creatures that no-one remembered and few
were interested in.

It was just then, surrounded by his workmates drinking stale wine that
the discussion turned to the weather. The winter was gathering momentum
and soon the ground would freeze up and the excavation would have to be
closed till the temperatures allowed digging again, which couldn`t be
expected until spring. Isaac was aware of this fact and with every day
being palpably colder then the previous one he new the time was
shortening. Yet he secretly feared the break the winter would provide him
with because the dig had become for him a life-nourishing substance.

⬦

The snow was falling heavily from the torn blankets overhead and Isaac`s
ears were prickling with cold as he walked to work. He knew what was
coming but there was something strong and unyielding in him that didn`t
want to comply with the idea.

Isaac entered the square of the Municipal House and the strong wind which
had been persistently blowing for the last three days suddenly changed
into a gale. Isaac ran for the shelter of the archway but when he got
closer he realized with sheer terror that the gate was shut. There was a
sign on the door painted in red letters and although not saying much,
still telling enough: "Excavation Closed!"

Isaac`s heart started to beat violently and his head filled with blood
that deafened his hearing. The gale was growing strong and there was a
roar of a thunder. Isaac was seized with panic. A snow-storm was coming,
yet it wasn`t the storm that Isaac was scared of. Suddenly he saw
something black and ominous sweep over his head. He looked up but there
was nothing there. The sky was a furious combat of the elements but there
seemed to be another battle going on. Isaac felt suddenly hot and dizzy,
his breath was heavy and stertorous . Panic came first and with panic his
state of mind worsened even more. He was losing control over what was
going on both inside him and without. He started to run.

⬦

125 pages the whole text..

---Page 336------------

On the Count of

Three

Isaac woke up. He was lying in a white bed which definitely wasn't his own. He tried to focus on the room beyond the boundaries of the blanket. It proved difficult at first, but even before he managed to take in the whole room an atrocious thought occurred to him and he knew where he was. The room had three other beds but they were empty and didn't bear any marks of recent occupation. The equipment was poor. One wardrobe and a bedside table next to each bed. The room was oblong rather then square, with a door on one side and a single window on the other. Close to the window there was a low table with three chairs neatly tucked under. One was apparently missing. The window was barred. Isaac tried to sit up but when he did so the whole room started spinning violently and the memories of the previous morning's terror flushed into his head like an icy cold water. He laid himself down again and with an anxiety bordering on fear and breathing heavily he coiled himself up into his blanket. The Mac Glee's Lunatic Asylum was well known throughout the city and Isaac for the first time in his life realized that this wasn't fun anymore. He was raving mad or maybe had been for who knows how long, and now he had a proof of it wherever he looked. Every piece of furniture, every blanket, pillow or a speck on the carpet talked of lunacy and they were long stories, and albeit Isaac couldn't know it he felt they were hideous tales of sorrow, confusion and pain.

Isaac fell asleep and when he woke up he thought he was hallucinating again. His mother was standing over him smiling with a smile that made you pity her: "You are ill, Isaac. You are very ill, my son," She said.

Isaac rubbed his eyes and after realizing that the surreal mirage was real he asked: "How long do you have to stay here, mom?" His mother looked at him properly. "But poor child, you must be hallucinating from the pills they give you. It's not me who is staying, it is you. I'll have to talk to the nurses about your medication." She turned to the door and looking as if she were about to leave she left. "A bit of rest might do her well, though." Isaac thought and the door closed with a clack.

It took three whole days before Isaac ventured out of his apartment. It wasn't as much out of curiosity but that he desperately needed a cigarette. His condition was poor and doctors had been working day and night to pigeonhole the strange aspects of his disorder. It didn't take Isaac long to find a smoking-room. It was small with nothing but one chair and a tiny bay window blowing in an icy cold draught, and Isaac dressed just in his dirty underwear and shaking violently from cold and weakness smoked three fags in a row to make up for the time he had been indisposed.

The corridor was swimming slightly when he reentered the hall and Isaac was about to return to his bed because his mind was slowly beginning to play tricks on him again. He was about to set off for his lodging when he noticed a figure about 20 feet away pacing the tiled floor as if waiting for something. "Waiting to get better I guess," he thought as he aimed his steps towards the inmate. "What else can one do in this shithole."

"I want to be a woman," said the boy looking dejectedly at Isaac and he reassumed pacing the corridor. He must have been waiting there long for he developed the habit of not treading on the red tiles which in a curious manner criss-crossed the floor. Isaac might have well lost his mind, yet he certainly haven't lost his common sense. "A woman, aye?" he repeated, measuring the bony lad with a tinge of disgust. The boy explained: "You see, I can`t stand

125 pages the whole text..

---Page 337------------

On the Count of

Three

being in this body any more. I want to be a beautiful bitch just as you
are." But that was enough for Isaac`s gentle nature. His manhood was
endangered and he became afraid of the boy. "Well, I`d better be getting
on," he blabbered, slightly ashamed and angry for the insult he had just
received and quickened his steps towards his room.

✧

The fifth day of his institutionalization came and Isaac was feeling no
better. He would spend most of the day in bed, involuntary indulging in
thoughts that left him shaking for hours with fear and desperation. When
he got better, a state which usually did not last longer than half an
hour, he would go outside the ward and stroll in a small car-park in
front of the building, because the snow was scooped away there into huge
piles on the margins, and there was no danger of his slippers getting
wet. The Mac Glee`s was an institution of colossal dimensions. It had
many buildings scattered on an area the size of the Main Airport. In
times of old it even had its own sugar refinery which Isaac could still
see out of his window looming in its dilapidation on the horizon and, as
many inmates thought, haunted by the ghosts of lunatics who never found
their way into the world of the sane again.

Nevertheless, Isaac was not to stay alone in his room for ever. The week
had just gone by and awakened from a strange dream about a pink, winged
elephant hovering above hills he had never seen and accosting him in a
language he did not understand, Isaac heavily crawled out of his bed.
There was a slight commotion behind the door and he wondered what for
pink elephants was going on.

Soon the door opened and in walked two men accompanied by a nurse. One of
the men looked in his fifties and he had a bald patch on his head. The
other was much younger, only slightly older than Isaac, and his head was
visible only when he took of his helmet, for he was dressed like a
warrior. The nurse kindly ushered them to their allocated beds and left
without a further ado. The older man took one of the chairs, pulled it up
to the window and, seating himself, he started staring into the garden.
The other took off his heavy, iron studded boots and, leaning a double-
handed sword against the wall, he laid himself down on his bed in his
chain-mail. "If you make a noise, I`m gonna kill you both." He said and
adjusted his pillow.

Isaac later learned that the man with a bald patch on his head wasn`t so
much a lunatic as a runaway from his wife and he had come to the madhouse
to have a breather. The other one was a different case, though. He was
twenty-six and had spent some years enacting historical battles. The
morning before he was flicked into the Bonnyluck`s Lunatic Asylum he
fought his last one. His army was winning with a flourish and there were
heavy casualties in the flanks of the opposing hordes. He felt himself
strong and on the pinnacle of his battle fury

125 pages the whole text..

--Page 338------------

On the Count of

Three

there was no stopping him. Yet something snapped. At first he thought
that his sword had broken, for he heard a terrible clang, but it wasn`t
his arm that was damaged, it was his brain. He was carried away on a
stretcher among the slain, crying like a baby.

✿

Christmas came and Mrs. Burton had come to visit her disordered son. "I
hear you are getting better, sonny, and they don`t give you so many pills
anymore." She enthused. "Undoubtedly it`s the company you have now," and
she gave the warrior who was just vehemently polishing his sword a
gratifying smile. Mrs. Burton was right in a way. Isaac had come to the
conclusion himself, yet it wasn`t for the positive influence they exerted
on him, but just the opposite. They were both so barking mad that Isaac
began to feel in their company quite normal.

"Mom?" Said Isaac.

"Yes, my dear?" Chirped Mrs. Burton. "I wanna go home." He whispered.

Christmas Eve at the Asylum was a dreary experience. Isaac had probably
never felt so depressed before. He was lying on his bed trying to read a
book. It was after some time that he was able to concentrate on something
again and it kept him going through the hardships he had to endure. Yes,
the hardships of Christmas in a madhouse. Every now and then he would get
distracted from his sulky reading by an outburst of glee emanating from
the common room downstairs. Everybody was there. Every single lunatic of
the ward along with the nurses was there. Yet booby parties were
something Isaac had never gone for. And so he would lie on his bed and
occasionally listen with cold terror to a cacophony of laughter which a
month ago he would have thought impossible for any human creature to be
capable of uttering. And then: "I`ll kill you all, you bastards." Then
laughter again.

On Boxing Day, they were all given a small present in the form of a
redoubled dosage and Isaac was obliged to hide the profusion of pills
under his tongue so that the nurse would think he had swallowed them, and
afterwards, under the excuse of an urgent call, go to the men`s room and
spit them all down the drain. The pills were strong and after swallowing
Isaac usually fell asleep in less then twenty minutes with something he
began himself to call a bloody-hefty-boulder syndrome.

Isaac was troubled. He was institutionalized because there was something
amiss with him and he knew that. Nevertheless, he had read too many books
to be so easily lulled into this dreamy world of the insane and feel
comfortable in it. While others were afraid of ghosts,

125 pages the whole text..

---Page 339------------

On the Count of

Three

voices, their own potential, or their wives, he was afraid for no reason whatsoever. "So what was the reason for it?" He wondered.

Isaac was even beginning to mull over the idea that, through so many meditative exercises and other pranks he had indulged in, his problem might have been purely spiritual. Yet that was an idea that gave him more anxiety than comfort because it would mean that his stay at the Bonnyluck`s Lunatic Asylum was of no avail and what he needed was a priest rather than a nurse.

His Mother came to visit him now and then and he beseeched her every time with a redoubled fervour. It was time to go home and albeit his mother wasn`t so perceptive as to understand, she was at last weak enough to succumb to his wishes. Of course the doctors protested, for they haven`t yet found the right drawer in which to put Isaac`s file and, by doing so, close an iron gate both over his case and person. They hurriedly met in the morning of Isaac`s departure and after half an hour of gurgling and murmuring and chirping issued a paper which said:

Name: Isaac Burton Status: single

Date of Birth: 11:11am.

Treated at the Mac Glee`s Psychiatric Clinic for the duration of three weeks. Symptoms: Panic fear, hallucination, excessive masturbation..

Conclusion: No traces of dystrophy found. Patient suffering from slight trichoschisis. Released for further observation.

Dr.: E. Jafar

◇

Isaac and his mother were sitting in a taxi cab. Isaac felt tired, exhausted after three weeks with loonies and head-cases far worse than his imagination could ever be able to embrace. Nevertheless, he felt only slightly better than he had done when he came to the Asylum. He still had his spells of excruciating fears and his thoughts were disarranged and seemed to lack the previous sensibility, so that when they finally got home he was glad to find his bed ready for a deep plunge into a sweet oblivion in which he was alternately to remain for the next three-quarters of a year.

Isaac spent the rest of the winter in a state not dissimilar to hibernation, subsisting on hot chocolate and

125 pages the whole text..

---Page 340------------

415

On the Count of

Three

cigarettes. Yet, his meager existence was not utterly without a purpose. Isaac soon discovered in his distorted thinking a great inspiration for poetry and through his prolific reading of English he soon began composing in it.

When the spring came Isaac would stroll in the garden of their house, for he did not dare to go further, and stretch his limbs and then sit for a while in the sunshine. Nevertheless, with the first marks of recuperation there came something that allowed Isaac little comfort. It was a grudge. A grudge against life which he thought himself betrayed by. Yet the wheel of time was in its constant spin and with the summer at hand there came a change into Isaac and he, albeit without knowing, began to attain something rare. And so it happened that where there used to be an inner strife and stifled anger came the soothing balm of humbleness and reconciliation. The exercise of the mind which he used to practice so heavily, and now felt, might have been the source of his present troubles he did not dare to execute. Nevertheless, in one of his old books dealing with spiritual techniques he found a simple exercise of breathing, and he would lie for hours without moving (not that he had been able to move anyway), and concentrate all his attention on his abdomen rising and falling with every taken breath. With earnestness he bettered his technique from day to day and soon was bathing in angelic states of mind he had never experienced nor ever dreamed of.

It was at the beginning of autumn that Isaac dared to cross the psychological boundary of his garden. In a tweed jacket and a shawl wrapped round his neck, for the chill of the season came early that year, he set off for a walk in the suburb. He knew his weaknesses now, but he was aware of his strengths, too. He was walking down a long allay hemmed with poplars and a scene of aimlessly ambling buddhas laughing at falling leaves came into his mind. A picture he once saw in a gallery of the Chinese art. His mastery of the English tongue was very good now, and without feeling superior, he knew that if he had graduated he might have now been a teacher in a school of certain reputation.

Isaac came home elated by the crisp air, although slightly tired because he didn`t feel quite himself yet, and there were still pills to take every morning and afternoon which made him feel heavy and dull. He didn`t like the pills because they tasted bitter and he hated them even more because he was dependent on them, and the idea of being dependent on something, save things he wanted to be dependent on, made him feel inferior and therefore ashamed. Isaac entered his room and laying himself on his bed began staring at the ceiling, a preoccupation which he had lately devoted a stupendous amount of time to. Nevertheless, it made his brain work and soon he was deep in thought. Isaac began weighing all pros and cons. He thought that even if he had the chance of becoming a teacher, the question still remained if he would be up to the job. Often he felt so weak he could barely get out of bed. Yet, he could not live this way for ever. Why, he could hardly stand the prospect of living this way till the end of the week. Isaac got up, his eyes aflare with decision. "I'll be a teacher

125 pages the whole text..

--Page 341------------

On the Count of

Three

even if it should break me". He said aloud. Then hearing his voice and the words uttered he started to laugh. He was already as broken as one can be, there was nothing to lose.

☼

"People live in a vacuum of their own experience. It is like a soap bubble around them through which they can see but their contact with the things without is superficial. People don`t like to touch these things because they hurt like every first contact with truth often does. Yet, when the bubble pops they are lost. Then it`s just the survival of the fittest." Isaac pondered. "But that`s Darwin." Retorted Jacob. "Then, I must be a Darwinist." answered Isaac.

Isaac was again visiting Jacob and they were deep in one of their versatile conversations. It had been nearly a month since Isaac sent out his curriculum vitas to nearly twenty schools all around the country including his own prep school which he had so bitterly disliked. Nevertheless, things were different now and Isaac, feeling himself changed, would accept any offer that came his way. Yet, with so many schools which he had bombarded with his measly life-story he had one more worry now. Every day, notwithstanding where he was, he could not be there after four o`clock, because that was the time when headmasters and

headmistresses all around the country, inspired by his unequaled feats should call and keenly expect to talk to the marvelous teacher whom Isaac felt himself to be. And so it happened that supported by few and dissuaded by many Isaac nevertheless got what his heart yearned for.

It was slowly drawing on towards the second month of his disclosure to the outside world, and Isaac was lying on his bed. His spirits were low and he was beginning to ponder the possible folly of his enterprise just when the phone rang. Isaac rushed to the source of the jarring noise and picked up the receiver.

"Mr. Burton?" Inquired the voice on the other side.

"Aye, that`s me," answered Isaac, trying to pinpoint the familiar voice.

"This is the headmistress of the O`Jerkin`s High School. Isaac, it looks like we would need your help here. Anyway, long time no see."

Isaac hung up. It was more than he could bear at that moment. The name of the

headmistress was Mrs. Trakny and she used to teach him chemistry in his fifth grade. She once nearly poisoned him to death by forcing him to drink a flacon of hydrogen peroxide thinking it was distilled water. She wanted to show the class that albeit without taste the water was still potable. The situation took on a strange, surreal twist, when Isaac, after his face took on the colour of the surrounding walls, started puking a white, dense porridge that began burning into the linoleum. The phone rang again and Isaac picked it up.

"Did you hang up on me?" Asked Mrs. Tranky. "I`m not sure what you mean?" Lied Isaac.

"Never mind, so what do you think?" Mrs. Tranky was impatient. "I`m your man." Said Isaac and hung up.

125 pages the whole text..

--Page 342------------

On the Count of

Three

✧

This was definitely something that Isaac could not foresee. To return to the high school he used to study at as a boy but himself now be the one to be listened to and learned from seemed to him so far-fetched till it began to make sense. After all, most of the former staff would still be there and it might be to his advantage that they should know him from before. Nevertheless, it meant moving again into a small city and Isaac new that this time it wouldn`t be in his mother`s tow.

Next day he packed his things. His plan was to go a few days earlier before the appointed time of his acceptance among the school staff, and rent a flat. His mother was to provide him with enough money to pay the first month and then it would be up to Isaac to take care after himself. Isaac definitely could not hope for earning a fortune because his qualification was low and he wasn`t properly educated, but he was prepared for the life of poverty. Moreover, he found the prospect of living in squalor quite romantic. He set out the same day. There was a through train going from the main station and his mother in all her magnificent grotesqueness came to bid him a farewell. She was a wretched figure to look at and Isaac couldn`t utterly banish the idea that it might be the last time they saw each other. The train slowly glided out of the station and catching a last glimpse of his mother`s waving hand, Isaac seated himself and prayed for the trustworthiness of angels.

He arrived late at night and not knowing what to do he did nothing till the morning broke out. When the day was up, Isaac, according to his mother`s advice, went to seek a man of a dubious character and dispositions who was to help him with a proper accommodation suitable for a high school teacher, and the man, being a magician in his own craft, put Isaac up in a beautiful flat the very next day.

The long expected day came and Isaac found himself standing in front of a big, brick-work building flooded with memories of old. Mrs. Tranky came to meet him at the main door and Isaac was swept away on a long and thorough tour round the school to her incessant and slightly irritating chatter.

The reason why they chose Isaac as a teacher was due to their former and very much praised Mrs. Tallow who had recently found out that her body was cheating on her with a malignant disease. She had a cancer and was bed-ridden in a local hospital, where she patiently awaited her last breath. Isaac felt slightly hard done by because they chose him as a quick substitute, but when he found out that Mrs. Tallow used to teach the older students and that he was to take up her classes, he was delighted.

Isaac was ushered into a closet that would serve as a room where he should prepare for his lessons and which he was to share with a woman he took an immediate liking to. She was a divorced female of forty named Shelly Crotz and, although in her relative prime in comparison to the rest of the school assemblage, her face was lined like a geological map of the Rocky Mountains. Yet, she had a sarcastic and slightly bitter humour which Isaac later

125 pages the whole text..

--Page 343------------

On the Count of

Three

found a great source of his amusement, and when she greeted him in a cordial manner Isaac new that they would be a team. Papers were signed and without much ado Isaac was released for the day with a timetable which showed the weakly arrangement of his classes. He woke the next morning feeling slightly nauseous. He didn't expect to be so nervous on the first day of his new job. His flat was situated in a relative proximity to the school and he decided to walk the distance which later showed itself to be no more than two miles. He got up early that day and being arrived to the school at seven o`clock, which was an hour earlier that the first lessons were scheduled for, he opened the door and climbed two flights of stairs to his closet. He made himself a coffee which had always had a soothing effect morning`s ravaged thoughts and, having placated himself, and with nothing to do, he did nothing. Shelly arrived half an hour later cursing about something he did not understand and later finding it belonged to her morning ritual, he soon ceased to be troubled by it whatsoever. Yet, sometimes even he, inspired by such morning profanity, fell a-cursing without really knowing why or whom.

Isaac browsed through the textbooks from which he was supposed to teach and, finding them trivial, he closed them again. He knew that his English was relatively good, and that in fact, it was much better than any other teacher`s at O`Jerkin`s High, and Isaac, soon discovering he had no rival, was obliged to feel superior.

A bell rang, announcing the beginning of a first lesson. Isaac took his textbooks and went to seek a door bearing the sign VIII. B which indicated the year the particular class was in and a group into which the year was divided. He found the door and without much ado entered the classroom. As soon as he walked in he walked out again and with a nervous twitch in his left eye he surveyed the sign with his healthy one. Then, finding there was nothing wrong with the plaque he took a deep breath and reentered. The classroom was full of giant teenagers with big heads and spotted faces emanating piercing shrieks and laughter. Two boys on the left side were just working on an inevitable destruction of one desk-table and when Isaac told them to stop, sparing him a scolding look with their dull, cow-like eyes, they only redoubled their effort. There was a girl wildly kissing with her classmate in the back row and Isaac when seeing the orgy was seized with such embarrassment and rage that all his rehearsed speeches were taken out of his mind.

"For my mother`s foul breath , shut the fuck up!" Isaac cried in English, himself surprised at the strength of his voice and the despair he felt. Yet the blasphemous phrase was like a soothing balm that descended on the classroom. All noise and movement ceased and in the blinking of his twitching eye Isaac was a witness to a change so sudden, that he couldn't but feel proud. Then one boy breaking the graveyard silence spoke, expressing the opinion of many:

"Can you teach us those things?" He dared. "Like what?" Asked Isaac, slightly confused. "You know, swearing and things like that."

125 pages the whole text..

On the Count of

Three

"Oh, sure. And many more," Said Isaac, beaming at the class.

Isaac gained the guys and he knew it. He opened a textbook on the page 12 and started teaching. In 45 minutes the bell rang, announcing the end of the lesson, and to the general acclaim of all present, Isaac left the classroom to return to his closet. There was just a ten-minute break between each class and when he slumped into his chair next to Shelly he already felt tired. Soon the bell rang again and standing up Isaac set off a-scouring the school for VI.A.

The Latin number six indicated that the pupils couldn't be more than twelve years old, which was the youngest class Isaac was to teach, yet he dreaded now what was to come. He found a door bearing the desired number in the basement and, entering a room devoid of sun, he was exposed to a spectacle beyond his wildest dreams. The classroom was in an uproar. The little people, as Isaac later began to refer to them, hearing that they were to have a new teacher, had gotten so excited that they decided to flood the classroom by the means of a wet- sponge fight while shrieking in their high-pitched voices which Isaac heard resounding in his ears for the rest of the day.

"For my mother's foul wrath…," blurted Isaac, but it was of no avail. He immediately understood that different rules applied down here. Isaac tried to resort to a calmer treatment of the situation. He positioned himself in the middle of the classroom and with mild gestures and placating words tried to restore order. He watched with sheer terror as some of the children fell on the floor and began crawling to his feet. He was just about to escape behind his teaching desk just when a peculiar occurrence offered itself to his sight. A bizarre creature in the shape of a small boy appeared out of nowhere demanding of Isaac to check the reference- book, while handing him the thick volume with a stupid smile that showed his braces, through which he was drooling words like through a sieve.

Isaac took the book with disgust and seating himself behind his desk, he decided to write down the absentees, even if it was the only thing he should do that day. Just at that time Mrs. Tranky was passing by in the corridor and hearing the commotion she knocked at the door of Isaac's classroom and entered.

"Is everything alright?" She asked, looking slightly alarmed at the sight in front of her.

"Everything is in a perfect order. We're just playing a small game on worms and beetles," answered Isaac, hiding his despair behind a forced grin.

The bell rang and Isaac could not but haste to his cubicle to have at least a sip of cold coffee

and a moment in which he would not be bothered by anyone.

"Tell me nothing, they are terrible." Said Shelly when she saw Isaac's pained expression.

" Are you sure you`re ok? Those `ittle beasts can make your day a hell on earth," she said, her forehead like a freshly ploughed field.

"It`s alright, I`m nearly half through the day," answered Isaac, half speaking to himself. But alright it wasn`t. At least not for the next month and the month after that.

Every day when Isaac got up he felt he was waging a war against something that was stronger than him. Yet his persistence was unassailable and with the beginning of the third month there came a change. It wasn`t a change in the weather, although winter was again

125 pages the whole text..

--Page 345------------

On the Count of

Three

slowly closing its cold hand on the country. It was an experience which
Isaac began to gain as a reward for his stubborn perseverance. And with
experience there came easiness into the way Isaac led his classes, an
easiness which the students began themselves to feel like they still felt
the milk of their mothers` breasts that not so long ago had nourished
their urine. Isaac`s gait had changed, and he now walked with a dignity
that pretends nothing, but is natal to the one who knows. His students
began to like him and where there used to be discord and confusion in his
lessons, harmony and silence now ruled with its supreme hand.

"They know fuck all." Cried Shelly who was correcting a week old tests
for her next class. Shelly Crotz was a good person but suffered bitterly
for her Polish name because for most of the school she came under the
general nickname of Mrs. Crotch. Even Isaac wasn`t spared, because most
of his students called him "Icy". Yet he didn`t mind, for he thought it
gave a certain strength to his English name Burton.

"How much time do we have?" Asked Shelly, slightly disconcerted from the
amount of tests she had yet to correct. Isaac looked at his watch, the
break had begun only recently.

"Well, take it like this. We have approximately six hundred seconds of
freedom," answered Isaac and gave Shelly an enigmatic look that slightly
disconcerted her. His favourite class was VII.C, and he was browsing
through some papers to see what they would be doing this time.

He acquired a definite calmness in his manner, his movements were slower,
and whatever he did was with a certain assurance. The bell rang.

The winter came and went, and spring was blooming crazily in its manifold
colours of joy. Yet, perhaps with everything germinating and blossoming,
there was something that began to grow in the deep recesses of Isaac`s
mind too, and what had at first been a hardly perceptible feeling, which
Isaac felt like a prod in his sleep, culminated into a crystal-clear
idea. Isaac was a different man to the one his few friends and even he
himself used to know. Every day he practised the exercise of the mind and
what used to be a carefree game had grown into a real desire for
spirituality. Isaac wanted to travel, and travel far, at that.

The end of the year was coming and Mrs. Tallow was miraculously
recuperating from her lethal affliction. And although her near
resurrection from the dead was truly fabulous, it nevertheless meant that
Isaac was soon to lose his job, because Mrs. Tallow was firmly bend on
returning the very next year.

"Shelly?" Said Isaac.

"Yeah?" Shelly was preoccupied with shoveling a fifth spoonful of coffee
into her mug. "I`m leaving." He said.

"What`cha wanna do?" Shelly was unperturbed.

Isaac heard of a land far across the see where the grass was always green
and there were sheep and cows walking freely, and the hills had stories

to tell. "People speak some forgotten language there and I heard that if you learn it, you understand the speech of birds," finished Isaac. Shelly was sipping at the concoction she had prepared for herself and with the look of

125 pages the whole text..

--Page 346------------

On the Count of

Three

a withered flower was measuring Isaac. "Look" she said. "I`ve been working here for nine years and look at me. I`m a wreck". Isaac looked at her and saw that she was. Shelly bowed her head as if thinking and then straightening up she beamed at Isaac: "I say, go. Run away till there`s any time left".

The bell rang.

Isaac was walking to his flat. There was just a week till the end of the year. Classification had been closed and children, with the prospect of the summer holiday in front of them, were becoming rebellious. Isaac cared little, though. Even for him it was a big countdown. And so it was that they all counted together:

1,2,3,4,5,6,7… Isaac Burton boarded a flight to Ireland on the 28th June at 11:11 am.

125 pages the whole text..

--Page 347------------

On the Count of

Three

125 pages the whole text..

---Page 348------------

On the Count of

Three

125 pages the whole text..

--Page 349------------

On the Count of

Three

Ben Smith

Born in 1983 in Prague, Benjamin Schmidt studies a private school of Art in Čakovice, on the Capital's outskirts, after which he successfully enrolls for the FF,UK, Prague. An artist, folk-singer and translator; his first book offers a hillarious insight into the human life, and condition. Three other books on. And sull

Gallerymarvel s.com

223 Pages the whole text...
…...

125 pages the whole text..

--Page 350------------

Cofee and bun in a French Restaurant close to Picadilly. Small wire art for Charity sales (above right,above).

Current Europian art (abstract informel), French magasine with current personages and fashion.

Small politics – Europe.

Vaclav Havel – Democratic betterments, renewments and landscaping. Milos Zeman – Giving world prices, inclinations towards KSCM.

Suggestions for a new President with British ancestry (a great amount of small political Parties in Europe with different views). Marcus Olan – France

Mary Mc ' Guiness – Ireland.

Large scale demonstrations of youngsters in France and All Europian strikes of workers (influx from eastern Europe).

The all mother iped congregation on French squares (expressing disatisfaction) 2015. Death of a Czech Archaeologist in Egypt (brutal murder) 2016. Street – Europian tv sieris, X-factor – Europian young talents.

---Page 351------------

Antropology by Benjamin Schmidt (inspired by Amy Whiten), Sarah Davids, France. Above (Dry needle, etching).

Exhibition in

Prague (Benjamin

Schmidt) Big in Japan, Game. Possibilities of modern painting with concrete features (pencil drawing),

abstract realism like modern photography (Curator). After Amy Whiten (Jon

Cohen).Benjamin Schmidt says.

Possibilities of abstract

expressionism and hiperrealism merge into logical prognoses like Hooking up and other Beatnik prose. Problems of intertwining steel and hipermodern buildings in structure upon half-painted large frame paintings. Drawing in Archaeology and other propoundments. Left above. Photographic licence (Right Above).

---Page 352------------

Animating plasticine dinosaurs for Film and Documentary purposes can be done and performed with subtitles. Interesting aftershading can lead to bettering of primeavel mood. (Benjamin Schmidt - above)

Interesting aspects of plasticine

have been in use for

anropological and

palaeontologi cal purposes.

--Page 353------------

Auvertly shading and color better the aspects of animation for scientific purposes. Animating and

recuntstruction of plasticine is known from about 1985. (Voracious Iguanodon- Benjamin Schmidt)

Palaeontological Bursa (Market) in Prague Louvre, 2015. Adiana Dolejsi with her colleuges (one from Mongolia) and Benjamin Schmidt.

Stones from Iraq and Africa for Christie ' s 2017 Bursa (Market). Possibly on pedestals.

250 000 B.C. Glacial Maximum (by Steven Mithen).

Discovered by Benjamin Schmidt, after Steven Culbreth (El. Granada, Ca.), Prague 2017.

A still-life with lentils

(differing values of touching up on computer for editing) Quality print
for advertisments and museum purposes- Above) Benjamin Schmidt
(Litography)

Quality in shading and oncoming aftershadind in computers for a livelier
look. A better prognoses for understanding material and final vision).
BenjaminSchmidt (Drawing)

--Page 355------------

A special Pot (Benjamin Schmidt), Interesting merging of abstract expressionalism and hiperrealism. Possibilities of pottery-making and glazing with deep thic colors.

Celebration of Indian instruments and instruments studying at Cork, Ireland. Palaeontological excavations. Fla festival. Pottery making. Loom in Dingle (Allan.P)

--Page 356------------
----

www.ingramcontent.com/pod-product-compliance
Lightning Source LLC
Chambersburg PA
CBHW031814170526
45157CB00001B/51